Seize the
Daylight

Seize the Daylight

THE CURIOUS AND
CONTENTIOUS STORY OF

DAYLIGHT
SAVING TIME

DAVID PRERAU

THUNDER'S MOUTH PRESS
NEW YORK

SEIZE THE DAYLIGHT
THE CURIOUS AND CONTENTIOUS STORY OF DAYLIGHT SAVING TIME

Copyright © 2005 by David Prerau

AVALON
publishing group incorporated

Published by
Thunder's Mouth Press
An Imprint of Avalon Publishing Group, Inc.
245 West 17th Street, 11th Floor
New York, NY 10011

First printing April 2005

First trade paperback edition 2006

Page 237 represents an extension of this copyright page.

Library of Congress Cataloging-in-Publication Data is available.

ISBN: 1-56025-796-2

9 8 7 6 5 4 3 2 1

Book design by Maria E. Torres
Printed in the United States
Distributed by Publishers Group West

For Gail and Michael

Contents

HORAS NON NUMERO NISI ÆSTIVAS

(*I mark only the summer hours*)

—Inscription on the William Willett Memorial,
Petts Woods, Kent, England

Benjamin Franklin Is Awakened Early

And the best of all ways
To lengthen our days
Is to steal a few hours from the night
—Thomas Moore, "The Young May Moon"

Benjamin Franklin conceived of it. Sir Arthur Conan Doyle endorsed it. Winston Churchill campaigned for it. Kaiser Wilhelm first employed it. Woodrow Wilson and Franklin Roosevelt went to war with it, and, more recently, the United States fought an energy crisis with it. For several months each year, for better or worse, it affects vast numbers of people throughout the world. And for one hundred years it has been a subject of recurring controversy in the United States, Britain, and dozens of other countries. But to trace the beginnings of daylight saving time, we must look first to Paris.

Benjamin Franklin was astonished.

"An accidental sudden noise waked me about six in the morning," he wrote in a whimsical letter to the *Journal de Paris*, "when I was surprised to find my room filled with light. I imagined at first that a number of lamps had been brought into the room; but rubbing my eyes I perceived the light came in at the windows." The year was 1784, and the seventy-eight-year-old Franklin—statesman, author, and scientist—was living in Paris while serving as the American minister to France. His attendant had forgotten to close the shutters the previous evening, and when Franklin saw the sunlight streaming through his windows, he checked his watch. It was just six o'clock in the morning.

"Still thinking it something extraordinary that the sun should rise so early," Franklin continued, "I looked into the almanac, where I found it to be the hour given for the sun's rising on that day. [Those] who with me have never seen any signs of sunshine before noon, and seldom regard the astronomical part of the almanac, will be as much astonished as I was, when they hear of its rising so early; and especially when I assure them *that it gives light as soon as it rises.* I am convinced of this. I am certain of my fact. One cannot be more certain of any fact. I saw it with my own eyes. And, having repeated this observation the three following mornings, I found always precisely the same result."

Franklin's "discovery" led to "several serious and important reflections." He realized that had he risen at noon as usual, he would have slept through six hours of sunlight. In exchange, he would have been up six additional hours that evening by candlelight. Since candlelight was much more expensive than sunlight, Franklin's "love of economy" induced him to "muster up what little arithmetic" he had mastered to calculate how much the city of Paris could save by using sunshine instead of candles.

For the six months between March 20 and September 20, Franklin estimated that on average Parisians would sleep seven hours after sunrise, and therefore could save seven hours of candlelight if they rose with the sun. Thus, he computed:

Number of nights from March 20 to September 20:	183
Hours each night when candles are burned:	7
Total hours (183 x 7):	1,281
Families in Paris:	100,000
Total hours in Paris spent by candlelight:	128,100,000
Total weight of candles consumed, at half a pound of wax and tallow per hour:	64,050,000 pounds
Total cost, at 30 sols per pound:	96,075,000 livres tournois

Ninety-six million *livres tournois* is the equivalent of about $200 million today—"an immense sum that the city of Paris might save every year by the economy of using sunshine instead of candles!" Moreover, Franklin added, "You may observe that I have calculated upon only one half of the year, and much may be saved in the other, though the days are shorter. Besides, the immense stock of wax and tallow left unconsumed during the summer will probably make candles much cheaper for the ensuing winter."

Although Franklin wrote in what one historian termed "a happy combination of humor and prudent instruction," he had obviously given the subject much thought. In fact, the germ of his idea can be traced back many years. In 1757 he made a similar observation in London: "In the summer, when the days are long... in walking thro' the Strand and Fleet-street one morning at seven o'clock, I observ'd there was not one shop open, tho' it had been daylight and the sun up above three hours; the inhabitants of London chusing voluntarily to live much by candlelight, and sleep by sunshine, and yet often complain, a little absurdly, of the duty on candles and the high price of tallow."

Although Franklin quite intentionally overstated the total savings by assuming that all Parisians slept until noon, he was serious about the underlying principle. He concluded: "It is impossible that so sensible a people, under such circumstances, should have lived so long by the smokey, unwholesome, and enormously expensive light of candles, if they had really known that they might have had as much pure light of the sun for nothing."

To remedy this waste of both sunlight and candles, Franklin became the first proponent of government action to alter the hours of human activity to make the best use of daylight. Continuing in the whimsical yet practical vein of his letter to the *Journal*, he put forward a four-pronged "Economical Project":

1. Let a tax be laid . . . on every window that is provided with shutters to keep out the light of the sun.

2. Let . . . no family be permitted to be supplied with more than one pound of candles per week.

3. Let guards be posted to stop all coaches, etc. in the streets after sunset. . . .

4. Every morning, as soon as the sun rises, let all the bells in every church be set ringing; and if that is not sufficient, let cannon be fired in every street, to wake the sluggards effectively, and make them open their eyes to their true interest.

This prudent plan was certainly in keeping with the man who, in *Poor Richard's Almanack*, had written "Early to bed and early to rise makes a man healthy, wealthy, and wise" (though obviously he had not always practiced what he preached). Fortunately for late-sleeping Parisians, Franklin's Economical Project was never put into effect.

DAYLIGHT SAVING TIME

Like Franklin's morning cannon firings, the basic goal of daylight saving time (DST) is to change the hours of human activity to make the best use of daylight. Rather than waking everyone early, however, as Franklin proposed, daylight saving time shifts the official clock time to provide extra daylight in the early evening in exchange for less daylight in the early morning. The clock is usually moved one hour forward in spring and back in fall, giving rise to the mnemonic phrase "Spring Forward, Fall Back." Although DST doesn't actually *add* daylight, it does provide more *usable* hours of daylight. In that sense, DST "saves" daylight—especially during the months when the sun rises before most human activity begins. "Daylight saving time" is considered to be the correct term for this clock-altering process, since it refers to a time for saving daylight, but "daylight *savings* time" is also commonly used.

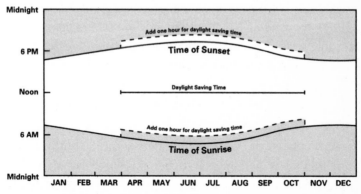

Daylight saving time shifts sunrise and sunset one hour later from spring through autumn.

Consider a summer day in Philadelphia, Denver, Naples, or Istanbul: four cities that, because of their geographic positions, have about the same sunrise and sunset times as each other on every day of the year (as in the preceding chart). On July 1, for instance, standard time sunrise is 4:35 A.M. and sunset is 7:33 P.M. Thus a worker who awakes at 7 A.M. and returns home at

6:30 P.M. would have slept through about two and a half hours of sunlight and would have only one hour of sunlight to utilize after returning home. However, with daylight saving time in effect, both sunrise and sunset are shifted one hour later— sunrise to 5:35 A.M., sunset to 8:33 P.M.—and one of the "wasted" early morning hours of sunlight is recaptured. Those Philadelphians, Denverites, Neapolitans, or Istanbullus now have an additional hour of sunshine after work with which to do whatever they please.

The goal of daylight saving time—to use daylight to its maximum advantage—is generally recognized to be of universal benefit. Nevertheless DST has been surprisingly controversial. Throughout its long and contentious history, daylight saving time has had an impact on a wide variety of often unexpected areas, from energy conservation, agriculture, and transportation to recreation, street crime, television schedules, voter turnout, gardening, schoolchildren, vehicular accidents, the workings of the stock exchange, and even the inheritance rights of twins.

From Ben Franklin's era until today, the story of daylight saving time has been a fascinating and sometimes bizarre amalgam of colorful personalities and serious technical issues, purported costs and perceived benefits, agendas of interest groups and policies of governments, pushes for uniformity and pulls for regional differences. In fact, the sunlight that stole through Ben Franklin's window on that Paris morning has preoccupied the thoughts of more than a few scientific minds, and the energies of numerous political leaders, in a host of intriguing and unexpected ways in the two hundred–plus years since it awakened Franklin.

Chapter One
The Waste of Daylight

O, for an engine to keep back all clocks.
—Ben Jonson, *The New Inn*

*A*s a beautiful morning dawned in the small town of Chislehurst, England, just south of London, in the summer of 1905, William Willett was up early for a horseback ride before breakfast, as was his custom. Willett was a well-known designer and builder of stately houses. Born in 1856, he had risen to the head of a leading firm of builders and was also a fellow of the Royal Astronomical Society. The houses he had built in the better neighborhoods of London were so widely recognized that the term *Willett-built* was a standing recommendation. He was also an ardent horseman, and he wouldn't miss his daily early-morning ride. Wasn't seven in the morning the most joyous hour of the day?

Riding across Chislehurst Common after his usual canter through the woods and over the downs, Willett observed that, as was frequently the case, he was the only one up and about. Like Franklin over a century earlier, he had long since recognized that during a large part of the year the early-morning daylight was scarcely utilized. Typically he saw no one in the early-morning light of spring and summer except the occasional postman, milkman, chimney sweep, or laborer, and the shutters and blinds of the houses he passed were all closed tightly to keep out the sun. Willett found it hard to understand why his countrymen were wasting the best part of a summer day. Could Englishmen, he wondered, ever be convinced to wake up earlier?

William Willett, daylight saving time's first champion.

Willett also recognized that most people had limited time after work to enjoy the outdoors before the setting of the sun. Children had to curtail their play, and outdoor enthusiasts such as gardeners were forced to stop their activities when the daylight faded. As a passionate golfer (another in his wide range of interests), Willett particularly regretted having to cut short his round when it got dark in the early evening.

Willett always had a special interest in light. His house designs emphasized the use of light and open space to such an extent that he was said to have "a love affair with daylight." The Willett design philosophy, as described by the trade publication *The Builder,* included this important tenet: "No matter how many little points in the exterior might be improved by the sacrifice of a little light somewhere inside . . . the sacrifice was never permitted, for he looked on light, especially sunlight, as all important."

During his morning ride, as Willett continued to reflect on the distressing waste of daylight, a revolutionary idea came to him: shifting the clocks forward to save daylight. This would allow some of the early morning's wasted sunlight to be utilized in the evening, and yet would not change anyone's normal waking hour. Thus was born the idea of daylight saving time.

Having conceived the notion of clock shifting, Willett grappled for the next two years with the particulars of a daylight saving plan, giving consideration to many possible benefits and many potential objections. Eventually he arrived at a comprehensive proposal, which in July 1907 he detailed in a pamphlet, *The Waste of Daylight.* He distributed *The Waste of Daylight* widely, at his own expense, and with its publication he opened an all-out campaign for the adoption of summer daylight saving time in the United Kingdom.

HOW MANY ADVANTAGES WOULD BE GAINED BY ALL

Willett began *The Waste of Daylight* by describing the situation he had observed: "The clear bright light of early morning, during spring and summer months, is so seldom seen or used. Nevertheless, standard time remains so fixed, that for nearly half the year the sun shines upon the land, for several hours a day, while we are asleep; and is rapidly nearing the horizon, having already passed its western limit, when we reach home after the work of the day is over. Under the most favourable circumstances, there then remains only a brief spell of declining daylight, in which to spend the short period of leisure at our disposal.

"Now, if some of the hours of wasted sunlight in the spring, summer, and autumn could be withdrawn from the beginning, and added to the end of the day," Willett continued, "how many advantages would be gained by all, and in particular by those who spend in the open air, when light permits them to do so, whatever time they have at their command, after the duties of the day have been discharged!"

Willett proposed a "simple expedient" to secure these advantages. His plan called for setting the clock forward by twenty minutes at 2 A.M. on each of four Sundays in April, for a total advance of one hour and twenty minutes, and setting it back by twenty minutes at 2 A.M. on each of four Sundays in September. "If we will reduce the length of four Sundays by twenty minutes, a loss of which practically no one would be conscious, we shall have eighty minutes more daylight after 6 P.M. every day during May, June, July, and August, and an average of forty-five minutes more every day during April and September." These four "short" Sundays of 23 hours 40 minutes and four "long" Sundays of 24 hours 20 minutes were "the whole cost of the scheme. We lose nothing, and gain substantially. . . . The advantages aimed at follow automatically, without any trouble whatever."

To silence potential naysayers, Willett pointed out that travelers by sea had easily accommodated themselves to the frequent alterations of time on board ship. They simply adjusted their watches and then went about their business. Closer to home, when travelers crossed the Irish Channel to Ireland, which followed Dublin Mean Time—twenty-five minutes behind Greenwich Mean Time—they reset their watches twenty-five minutes backward or forward and, Willett contended, thought nothing of it. He reflected that it is less trouble to move the hands of a watch than to wind it up, an operation most people of that era performed 365 times a year. As to concerns about making a variation from standard time, he observed that in Australia, South Africa, India, and other places around the world, the local time had been set to deviate from standard time by a fixed amount, such as one-half hour. (He failed to mention, however, that there was no country where clock time changed several times per year, as it would under his plan.)

His proposal, Willett continued, would provide more time that could be spent in exercise in the open air. "The brief period of daylight now at our disposal between the hours of work and sleep is frequently insufficient for most forms of outdoor recreation, but the daily addition of one hour and twenty minutes after 6 P.M. will multiply, several times, the usefulness of that which we already have." The benefits of parks and open spaces could be doubled and, with some interesting foresight, Willett added, "The nation may, some day, have cause to be thankful that, by this means, opportunities for rifle practice will have been created."

THE·WASTE
OF·DAYLIGHT

WILLIAM·WILLETT
F·R·A·S·

Price·3ᵈ   Copyright.

Willett's famous pamphlet.

Echoing Franklin's appeals to economy of 123 years earlier, Willett argued, "Everyone, rich and poor alike, will find their ordinary expenditure on electric light, gas, oil, and candles considerably reduced for nearly six months in every year." And, like Franklin, though without any exaggeration, he demonstrated the overall financial benefit of his scheme:

Assumed cost of artificial light for each person: 0.1 pence per hour

Additional hours of daylight per year: 210

Cost savings per person per year:	1 shilling 9 pence
Total population of Great Britain and Ireland:	43,660,000
Total estimated cost saving per year:	£3,820,250
Deduct, to meet all possible objections, say ⅓:	£1,273,416
Minimum net annual saving:	£2,546,834

The £2 million saved annually would be the equivalent of over £100 million, or $200 million, today.

The use of more natural light would provide several additional benefits. Less coal would be required for the production of gas and electric lighting, increasing the longevity of coal supplies. Much less smoke would "defile the air." Eyestrain would be reduced. Furthermore, the additional hours spent outdoors would spark an improvement in overall health. "Against our ever-besieging enemy, disease, light and fresh air act as guards in our defence, and when the conflict is close, supply us with the most effective weapons with which to overcome the invader." Beyond the physical benefits, more daylight would also boost the nation's mental health and morale. "While daylight surrounds us, cheerfulness reigns, anxieties press less heavily, and courage is bred for the struggle of life." Under his proposal, "every year of life will be leavened with additional gaiety."

As to potential objections, Willett first addressed himself to the effect of his proposal on train schedules, stating "With one exception, all trains will run in accordance with existing time tables." The few trains running at 2 A.M. on the four Sundays in April would arrive twenty minutes late, but both passengers and officials would know this before each journey commenced. The only significant effect would be on trains connecting via steamer to trains in continental Europe. These trains would need three special timetables: one for April, one for May through August, and one for September, but it would be worth the trouble: "Railway Companies will not only find ample compensation in reduced expenditure on artificial light on stations and in carriages, but as people are more ready to travel before than after sunset, increased passenger traffic." Both astronomy and navigation by international agreement used Greenwich Time, and so these two areas were specifically exempted from the time changes in his proposal.

Willett anticipated another possible objection: that daylight later in the evening might result in shortened sleep. He countered this by pointing out that most would actually sleep better, for more light in waking hours would leave more darkness for sleeping hours. For those who sleep late, "The advantage of not being awakened by the sun so early, as they sometimes are, will not be unappreciated." Willett countered the objection that the benefits achieved by shifting the clock could equally be obtained by early rising by asserting, "Leisure must follow, not precede, work, and compulsory earlier business hours are quite unattainable." Though some inconvenience might be experienced as a result of to the time difference between England and other countries, Willett contended that this loss of convenience would be slight. Furthermore, any difficulty would exist for only half the year, and ultimately they would disappear "when the advantages we shall have gained lead other nations to follow our example."

In conclusion, Willett extolled the cumulative benefits of his proposal: "On reaching the age of twenty-eight (without counting anything for six years of childhood), a man will have gained a whole year of daylight. At fifty he will have gained two years, at seventy-two—three years! Can any words be comprehensive enough to represent the cumulative effect of these additional hours of daylight . . . which are within our reach, to be had not only without price, but accompanied by a large saving in current expenditure year after year?" He ended with a call to action: Everyone wishing to obtain the extra hours of daylight "must do something. Out of nothing, nothing comes." He urged his readers to appeal to their friends and colleagues for support and, to urge their member of Parliament to support a bill to save daylight. "*Everyone* who wishes for light evenings MUST do this. No one can sit in an easy chair and climb a hill at the same time."

A "STARTLING" BILL IN PARLIAMENT

Robert Pearce, Member of Parliament (MP) for Leek, in Staffordshire, favored Willett's ideas as soon as he learned of them. He wrote to Willett, "The high merit of your proposal is the simple way of it, and the extraordinarily slight disturbance of existing things." Pearce soon drafted a bill embodying Willett's plan. Willett had never specifically used the phrase "daylight saving" in his pamphlet, but now the phrase became the name of Pearce's Daylight Saving Bill, which he introduced in the House of Commons on February 4, 1908.

"The purpose of the Bill," Pearce wrote, "was to bring the hours of work and pleasure nearer to the sunlight." Although today daylight saving time might seem a less than revolutionary idea, at the time *Scientific American* declared: "It is not often that a measure of such a startling character as the Daylight Saving Bill is introduced to the English House of Commons." By March, Willett had obtained the support of nearly two hundred members of Parliament, and a Select Committee, chaired by Sir Edward A. Sassoon, had scheduled hearings to discuss the merits of Willett's proposal.

Willett appeared before the Select Committee on April 7. That morning, he observed, the sun had risen at about 4:20

A.M., but most people had had breakfast at about nine o'clock. Five-eighths of what he called "the youth of the sun" had already gone before most of them had entered upon their duties. Willett presented to the committee many of the points he had made in *The Waste of Daylight*, using several sets of documents and diagrams to illustrate the value of saving daylight.

Several others who testified pointed to the practical problems of having many time changes in a year. The changing of

---

THE DAYLIGHT SAVING BILL

Be it enacted by the King's Most Excellent Majesty, by and with the advice and consent of Lords spiritual and temporal, and Commons, in this present Parliament assembled, and by the authority of the same as follows:—

From and after the passing of this Act

1. The hour between two o'clock and three o'clock in the morning of each of the first four Sundays in April in each year shall be a small hour consisting of forty minutes only but shall otherwise be reckoned a full hour for all intents and purposes whatsoever in the United Kingdom.
2. The hour between two o'clock and three o'clock in the morning of each of the first four Sundays in September in each year except 1908 shall be a long hour consisting of eighty minutes only but shall otherwise be reckoned a full hour for all intents and purposes whatsoever in the United Kingdom.

3. Greenwich Time as used for the purposes of astronomy and navigation shall not be affected by this Act.

4. The variation of time hereby established shall be known as British Time.

5. This Act shall apply to the United Kingdom of Great Britain and Ireland and shall be known as The Daylight Saving Act, 1908.

Robert Pearce's 1908 Daylight Saving Bill.

the clock forward in the spring in four steps of twenty minutes and back in the autumn in another four steps of twenty minutes was called by opponents a "policy of pin-pricks." A number of alternative schemes were proposed to the committee, including three steps of thirty minutes, three steps of twenty minutes, and two steps of thirty minutes. Representing the Post Office, Sir H. Babington Smith expressed the opinion that the total change should be one hour, not an hour and twenty minutes, and that, while two changes of thirty minutes might be acceptable, a single change of an hour was much preferred. Admiral E. R. Fremantle felt that one change was best and that it should be forty-five minutes, as an hour would make sunrises too late at Penzance, Cornwall, the westernmost town in England. J. F. S. Gooday of the Great Eastern Railway and other railroad executives favored the principle of the bill but preferred that only a single alteration of time be made in spring and autumn. They felt that if the clocks were moved forward on four successive weeks, there would be great confusion on the railroads. At each time alteration, every engineer, guard, and signalman would have to alter his watch, and if one of them forgot to do so, the consequences could be disastrous.

In further testimony it was pointed out that four small changes in time would cause undue public inconvenience and result in too much interference with the ordinary measurement of time by clocks and watches. Lord Avebury observed, with British understatement, that the eight annual time changes proposed in the bill were "rather numerous." Accepting the point, the committee decided, "for the sake of simplicity and general convenience," to modify the bill to a system with a single alteration of the clocks in the spring and a single setback in the autumn. Under this revised plan, the clocks would be set forward one hour at 2 A.M. on the third Sunday in April, and set back one hour at 2 A.M. on the third Sunday in September. Willett accepted this modification and fully supported the revised bill, reworking subsequent editions of *The Waste of Daylight* accordingly.

GLORIOUS SUNSHINE, WHICH COSTS US NOTHING

From the first publication of *The Waste of Daylight* to the introduction of the Daylight Saving Bill in Parliament to the ultimate resolution of the issue, strong opinions—both pro and con—were expressed by prominent people, publications, and organizations concerning the radical notion of reconfiguring the hours of the day.

Many members of Parliament voiced support for the daylight saving proposal, including four former or future prime ministers: Arthur J. Balfour, David Lloyd George, J. Ramsay MacDonald, and Winston Churchill. King Edward VII indicated that he, too, was favorably disposed toward the bill. The king had recognized the waste of morning daylight and had already taken royal action: For several years, to have more time for hunting, he had been creating his own small sphere of daylight saving time at his palace at Sandringham, and in later years at Windsor and Balmoral castles, by having all the clocks advanced thirty minutes.

Various constituencies had their say in support of the bill. "The medical case for the Bill is a strong one," declared Sir Thomas Barlow, president of the Royal College of Physicians. "An extra hour of daylight for all purposes, instead of artificial light, from the viewpoint of eyesight, and therefore health (because eyesight affects the health), would be a great national asset." G. H. Burford, president of the International Congress of Homeopathic Physicians, agreed: "Science teaches us, ever more impressively, how much daylight and sunshine mean for public health. The waste of the light of the morning hours is a flagrant squandering of that health." Dr. Robertson Day, a fellow of the Royal College of Surgeons, felt that the bill would be "healthfully beneficial," and that if exposed to more sunshine, people would suffer less from rickets, anemia, and eyestrain. The extra daylight would even benefit the blind, according to A. W. G. Ranger, chairman of the British and Foreign Blind Association: "To the small section of the community to which, as a blind man, I belong, the gain would be peculiarly joyous, and beneficial; for, to the blind, sunlight and fresh air are essential conditions precedent to good health, and really enjoyable life."

Businessmen such as C. F. Higginson, the manager of the National Bank, wrote that the suggested change "is one of infinite good, and appeals strongly to those, like myself, whose lives are spent in close and somewhat dark conditions of City life." (The City is London's financial district.) Lord Avebury mentioned that the bill not only would be a great convenience to merchants and bankers, but also, and even more important, would give more time to clerks for a game of cricket or other recreation. Richard Burbridge, managing director of Harrods Stores, said that his company made fourteen acres of recreation ground available to its four thousand employees, but they were unable to make much use of it on summer evenings. The directors of Harrods supported the bill so strongly that the company later printed and issued at its own expense fifty thousand copies of a pamphlet explaining and endorsing it.

Several representatives of railroad companies endorsed the bill, anticipating an increase in passenger traffic and a decrease in lighting costs for carriages and stations. Some expressed concern over trains that connected with steamers going to continental

Europe, and also mentioned the problem of mail from the Continent arriving an hour late. Others felt that connections with the Continent offered no serious difficulties. Sam Fay, general manager of the Great Central Railway, pointed out another advantage for the railroads: a far greater proportion of dangerous operations, such as shunting (switching freight cars from one track to another in a freight yard) would take place in daylight.

Among what today might be called "celebrity endorsements" was support from Sir Arthur Conan Doyle, well known then and now as the creator of Sherlock Holmes. "It seems very strange," he remarked, "that in the course of the world's history so obvious an improvement should never have been adopted." Daylight saving time "would make for the health and happiness of the majority of the community, and the next generation of Britishers would be the better for having had this extra hour of daylight in their childhood. . . . The objections are in the minority as compared with the advantages."

Summing up the case for daylight, as Benjamin Franklin had done more than a century earlier, Sir Robert Ball, probably the best-known astronomer of the time, wrote: "Which is the better for our waking hours, glorious sunshine, which costs us nothing, or expensive and incomparably less efficient artificial light? . . . The admirable scheme of Mr. Willett will rescue 210 hours of our waking life from the gloom of man's puny efforts at illumination, and substitute for it—sunbeams." He concluded: "Meridians were made for man, not man for meridians. Time must be regulated . . . to suit man's convenience."

THE OPPOSITION MOUNTS

Despite such strong support, Willett's proposal was not without its critics. The opposition was sizable, and controversy raged not only in Parliament but also in the press, as editorials and a series of long letters to the editor attacking and defending the bill appeared in newspapers and scientific journals. Many of the same arguments used during these initial debates have resurfaced each time the adoption or expansion of DST has been proposed, up to the present day.

Most scientists and astronomers were opposed to any tampering with time. Sir William H. M. Christie, the astronomer royal, declared that the scheme was simply special legislation for the benefit of late risers. Setting the hours of labor to make the best use of daylight should be done without "juggling with the measurement of time." Sir William Napier Shaw, director of the Meteorological Office, declared that "to alter our present mode of measuring time would be to kill a goose that lays a very valuable egg." He worried about the practical effects of adding or skipping an hour. These leaps in time would produce difficulties—for example, in the compilation of data from continuously recording meteorological instruments. The editors of the scientific journal *Nature* jibed, "It would be more reasonable to change the readings of a thermometer at a particular season than to alter the time shown on the clock, which is another scientific instrument." They wondered if perhaps another bill would be proposed "to increase the readings of thermometers by ten degrees during the winter months, so that 32° F shall be 42° F. One temperature can be called another just as easily as 2 A.M. can be expressed as 3 A.M.; but the change of name in neither case causes a change of condition."

Further, the transportation executives, scientists, businessmen, and others who had achieved a worldwide standard time zone system objected to the introduction of any irregularity. *Nature*'s assessment supported this view. "The advance from 'local' to the 'standard' time of today was a step well

thought out, and one that cannot be reversed by the introduction of a new and really nondescript time under the old name."

Some of the strongest opposition came from agricultural
interests. Farmers complained that they could not change their
daily schedule and start work an hour earlier just because the
numbering of the hours was changed. Several farm operations
could not possibly be performed earlier than they were at
present. For example, the harvesting of grass for hay could not
be done before the dew had disappeared, whatever the clock
said, because the reaper and binder machines would not work
unless the hay was dry. Other farm activities could not be done
until the cool of the evening, and thus farmers could not end
work an hour earlier, no matter what the clock said.

Merchants who conducted trade with continental Europe
felt daylight saving time might interfere with business. Those
who dealt telegraphically with the United States objected to
shortening the overlap of business hours, as there was a lively
exchange of cablegrams when business hours in Liverpool and
New York coincided. E. Satterthwaite, secretary of the London
Stock Exchange, said that the bill would "create a dislocation
of Stock Exchange business in the chief business centre of the
world." He thought the time changes would cause irritation in
other countries and lead to business drifting away from
London. He drew attention to the fact that the stock markets of
New York and London were open concurrently for only one
hour. If the bill passed, their hours would not overlap at all
during five months of the year.

A common argument was that even though making more
daylight available was a useful objective, it could be better
achieved by changing people's habits, not the clock. Anyone
who desired it could obtain the same effect without the need
for daylight saving time legislation, simply by waking up one

hour earlier. Businesses could easily shift their hours of operation. "If, for example, the Bank of England could be persuaded to open business at 9 A.M. instead of 10 A.M. from April 1 to September," said Sir David Gill, "no doubt all other banks and offices would follow suit, and if employers of labour would open their works an hour earlier in the spring and summer months the objects of the Bill would be in great part gained without difficulty and confusion." Some English employers had already moved the workday forward a half hour or hour in the summer. In a large warehouse on Queen Victoria Street, notices were posted with this message:

NO POLITICS IN BUSINESS.
WHY WAIT FOR PARLIAMENT TO ALTER BUSINESS HOURS?
THESE PREMISES WILL BE OPEN AT 8 A.M.
AND WILL CLOSE AT 5:30 P.M. FROM JUNE 1 TO OCTOBER 1.

Reinforcing their opinion that Willett's proposal was much ado about nothing, the unswerving editors at *Nature* magazine wrote dismissively, "All that is needed is for banks, places of business, and schools to open at an earlier hour during the summer months. . . . To introduce confusion into the whole system of time-reckoning because some people in cities have not sufficient strength of mind to make the best use of daylight hours would be to acknowledge that, as we cannot alter our national habits and customs, Acts are passed by which we pretend to change them while they remain the same."

As the debate became more heated, some opponents resorted to ridicule. "Will the cows give their milk earlier because of Mr. Willett? . . . Will the chickens know what time to go to bed?" Willett himself became the object of a great deal of derision; he was called everything from a faddist to a dangerous crank. Some mocked what they considered the artificiality of Willett's plan. "Suppose," scoffed Sir Herbert Stephen, "that the legislature thought it desirable that most men wear white hats, and accordingly enacted that every hat possessing the qualities which we now signify by the word black should . . . be called white. The result would be that we should mostly wear the same sort of hats as present, and that for official purposes the considerable majority of our hats would be statutorily white."

Astronomer Sir George Darwin (son of the eminent naturalist Charles Darwin) equated the DST plan with an experience he had as a little boy. Mortified at not being six feet tall, he marked his height "carefully on the playroom wall, and divided it into a scale of six feet, and the feet into inches." From then on he was six feet tall (and his little brother was five feet ten inches). In a letter to *The Times*, "H. B. D." stated, "Even Parliament cannot alter the relative positions of the sun and the earth. It can only order all the clocks in Great Britain to be wrong for five months in the year, so that young people may play more games." And a poem published in the humor magazine *Punch*, written "in apprehension of the Daylight Saving Bill," dismissed such benefits as extra daylight for playing cricket, and insisted that DST should not alter the time at which a servant prepares the morning tea:

> "Cricket" (the fanatics urge) and "economy,"
>> "Saving of gas"—do I care about that?
> Think of the charm of our childhood's astronomy,
>> Think of the soft and marsupial bat:
> Think of the authors of sonnets that ruminate
>> Under the stars by the silvery Thames;
> Think of the thousands of ads. that illuminate
>> London by night with electrical gems.

No, by the might of the Muses that foster us!
    Let them, advancing the hands of the clock,
Force on the masses a wholly preposterous
    System—but we will be firm as a rock.
Others, surprising the sun in his chariot
    Long ere their wont, may submissively delve,
We must demand of Eliza (or Harriet)
    Not to be called at eleven, but twelve.

Finally, *The Outlook* pointed out two practical difficulties that "have, with vulgar persistence, obtruded themselves. The person who now dines at 7:30, for instance—which is perhaps the average dining hour of the Londoner—will then be dining at 6:10, which is preposterously early, and will be altogether unfashionable. Moreover, there is one aspect which would fill London with horror. If, for instance, a man were going to a seven o'clock dinner, under the new arrangement of daylight he would appear on the streets of London in evening dress at 5:40, which would shake the British Empire to its foundations."

DST IN PARLIAMENT

After meeting thirteen times and hearing testimony from forty-two witnesses, the Select Committee reported favorably to Parliament on the Daylight Saving Bill. In its Special Report issued June 30, 1908, it defined the bill's objective as "promoting the earlier and more extended use and enjoyment of daylight" April through September. The committee found that this objective "is desirable and would benefit the community if it can be generally obtained." It further found that "the weight of the evidence submitted to the committee agrees with and supports this view, though there was divergence of opinion as to the best mode of accomplishing it."

The committee listed six principal benefits of the greater use of daylight:

1. To move the usual hours of work and leisure nearer to sunrise.
2. To promote the greater use of daylight for recreative purposes of all kinds.
3. To lessen the use of licensed houses.
4. To facilitate the training of the Territorial Forces.
5. To benefit the physique, general health, and welfare of all classes of the community.
6. To reduce the industrial, commercial, and domestic expenditure on artificial light.

In its report, the Select Committee answered many of the criticisms leveled at the bill, stating, for example, that the effect on business with Europe would be small compared to the overall benefit, and that the interference with American business could be minimized without serious loss, as the various parties could adapt themselves to the changes. Greenwich Mean Time would still be used for all scientific purposes, so that was not an issue. As to whether legislation was needed at all, the committee believed a single act establishing local time for the United Kingdom would be better than an act changing all the specific times prescribed by various bills, bylaws, and other rules.

But despite this positive report from the Select Committee, the opponents of DST stalled the bill. The real problem was that the bill did not have the support of Prime Minister Herbert Asquith. On July 8, 1908, Asquith told Parliament that his government had "no intention of giving facilities for the passing of the Daylight Saving Bill." Without Asquith's backing, the bill could make no further progress through the House of Commons.

Refusing to accept this reversal, Willett and his supporters tried again the following year, when MP Thomas W. Dobson introduced the 1909 Daylight Saving Bill. The motion was seconded by Sir Henry Norman, who sought to influence the proceedings by invoking an internationally admired personality. "Suppose Mr. Wilbur Wright could fly over this country at about five o'clock on a summer morning. He would see the most beautiful country, perhaps, in the whole world brilliantly lighted, fresh and wholesome, yet he would find that practically the whole population were sleeping behind curtains in their beds."

The bill was approved for committee hearings on March 5, 1909, and a second Select Committee eventually heard testimony from twenty-four witnesses. One witness was J. S. R. Phillips, editor of the *Yorkshire Post*, who explained to the committee that the bill lacked support among the members of the Newspaper Society because the measure would prevent early publication of some types of news; cricket matches, for instance, would not end early enough for evening newspapers to carry the final scores. Theater owners worried that their customers would not want to attend performances in broad daylight. The Associated Chambers of Agriculture, comprising over 150 agricultural organizations across the country, reiterated their opposition and declared that the proposed system would cause great inconveniences without any corresponding advantages.

At the close of the hearings, the committee rejected the bill by the margin of a single vote. Its final report, issued on August 26, 1909, presented two primary reasons for the committee's opposition: "the great diversity of opinion existing upon the proposals of the bill" and "the grave doubts which have been expressed as to whether the objects of the measure can be attained by legislation without giving rise, in cases involving important interests, to serious inconvenience."

AN EXTRA YAWN, AN EXTRA SNOOZE

Undaunted by this second defeat, Willett and his supporters continued pressing for a daylight saving bill. Willett marshaled the movement's backers and tirelessly worked to win new adherents, while continuing to spend large amounts of his personal fortune for the cause. He sent out hundreds of letters each year to influential people in all walks of life, and published several new editions of *The Waste of Daylight*, each naming more people who endorsed the proposal.

In 1911, Robert Pearce introduced yet another bill in Parliament proposing daylight saving time—now being called "summer time" or "summer season time." Willett organized a large public meeting in support of the bill, held at London's Guildhall on May 3. The lord mayor of London presided, and Winston Churchill, then home secretary and an ardent advocate of DST, was the principal speaker. Summoning the oratorical skills for which he would become renowned, Churchill delivered an eloquent and powerful speech of support. "An extra yawn one morning in the springtime, an extra snooze one night in the autumn is all that we ask in return for dazzling gifts. We borrow an hour one night in April; we pay it back with golden interest five months later." He projected both moderation and confidence. "If the change is not found to be beneficial, the experiment need never be repeated. But we are confident that advantages of a most substantial and important character would be immediately secured in almost every sphere and aspect of national life." Cheers rang out from the

hall. In conclusion, Churchill predicted that "a grateful pos-
terity, dwelling in a brighter and healthier world, would raise
statues in honour of Mr. Willett and decorate them with sun-
flowers on the longest day of the year."

But despite Churchill's eloquence and the cheers of the
Guildhall throng, the bill's opponents brought counterargu-
ments. *Nature* continued its strong opposition, declaring, "The
scheme is unworthy of the dignity of a great nation . . . and
would make us the laughing-stock of the enlightened people of
the world." When the 1911 Summer Season Time Bill was put
before Parliament, the results of the committee hearings of the
previous years were given considerable weight, and it did not
pass. Subsequent attempts in 1912 and 1913, led again by an
unrelenting Willett, met with a similar fate. When Willett's
supporters in Parliament introduced yet another summer time
bill in 1914, his *Waste of Daylight* was in its nineteenth edition;
it listed endorsements from influential persons and institutions,
including 285 members of the House of Commons; 59 members of the House of Lords; 685 city, town, and district councils;
82 chambers of commerce; 59 trade unions; and 438 business,
political, and other societies and associations. But all of this
support was unable to generate a parliamentary majority—the
bill was once again rejected.

While the United Kingdom continued to debate daylight
saving time, Willett's proposal was sparking interest around the
world. *The Waste of Daylight* was translated into several lan-
guages. At the 1914 International Congress of Chambers of
Commerce, in Paris, Willett gave a talk on daylight saving time
to delegates from thirty-seven countries. In France, the editor
of *La Petite République* wrote about the "advantages to be
derived from Mr. Willett's system," and the Associated Chambers of Commerce of the German Empire advocated DST for

Germany. Daylight saving time bills were introduced for Canada, New Zealand, and provinces of Australia, and although they garnered some support, ultimately none was approved.

On March 4, 1915, William Willett died at Chislehurst at the age of fifty-eight. His daughter Gertrude later wrote, "Above all, he loved sunlight, open spaces, and fresh air. This, and an untiring fund of energy, were his leading characteristics." Willett had put many years of undiminished enthusiasm and a large portion of his personal wealth into a valiant struggle to obtain passage of a daylight saving time bill—but he was never to see his revolutionary proposal come to fruition.

In the face of dogged opposition by one Parliament after another, it seemed that Willett's cause had been lost along with its galvanizing champion. Few could have imagined that his cherished idea would soon find renewed support, in a most unlikely quarter, under the pressure of world war.

Chapter Two
From Sun Time
to Standard Time

The clock, not the steam engine, is the key machine of the
modern industrial age.
—Lewis Mumford, *Technics and Civilization*

*W*illiam Willett's proposal for daylight saving time was
scorned by many for its artificiality. Yet Willett did not propose
"to change from natural time to artificial time," as Winston
Churchill told Parliament, "but only to substitute a convenient
standard of artificial time for an inconvenient standard of artificial time." Indeed, daylight saving was not the first but rather
the third artificial adjustment to natural sun-time. First there
was *mean time*, then *standard time*, and finally *daylight saving
time*. Each modification took timekeeping further from nature,
but each brought benefits that were generally considered to be
worth the change.

THE SIXTY-NINE-MINUTE HOUR
From the beginning of human history, ancient civilizations
sought to measure the passage of time during a day. The sun's
progress across the sky from horizon to horizon was used as the
timepiece. At some point early peoples came to understand

that the changing length of the shadow of an upright object, such as a person or a tree, was related to the time of day. Eventually they planted a vertical rod in the ground to track the sun's advance.

The ancient civilizations of Sumeria and Egypt needed a method of timekeeping to coordinate and regulate religious practices and the workdays of government officials. Vertical pillars such as obelisks were used to mark time. The shadow cast by the obelisk slowly shortened from sunrise to noon, and then lengthened again from noon to sunset. The time when the obelisk's shadow was shortest divided the daytime period into two parts. Obelisk shadows also allowed Egyptians to mark the annual solar cycle—from the summer solstice (the longest day of the year), around June 21, when the noontime shadow of the obelisk was at its shortest, to the winter solstice (the shortest day of the year), around December 21, when the noon shadow was at its longest. Later, radial markers were positioned on the ground at the base of the obelisk to indicate time subdivisions, based upon the changing angle of the shadow. This was also the concept of the sundial, which was in use in Egypt at least thirty-five hundred years ago.

The earliest systems for dividing the day used simple partitions. The Sumerians split the day into twelve periods they called *danna*, each consisting of thirty *ges*, mirroring their division of the year into twelve thirty-day months. The Egyptians divided a day into two sets of twelve equal units: The daylight period had twelve equal "hours of day," and the nighttime, twelve equal "hours of night."

The duration of daylight and darkness varies from day to day as a result of the tilt of the earth's axis with respect to the sun; consequently, the Egyptian daytime and nighttime hours would fluctuate in length throughout the year. Because of this they were referred to as *temporal hours,* meaning "unequal" or "temporary." A summer day would have twelve long temporal hours of day and twelve short temporal hours of night, and the reverse would be true in winter. The summer and winter solstices, with the longest and shortest daylight periods of the year, were thus the days with the maximum differences between daytime and nighttime hours. On the summer solstice day at Thebes, long the capital of ancient Egypt, a daytime temporal hour was about sixty-nine minutes and a nighttime hour lasted fifty-one minutes, a proportion approximately reversed on the winter solstice. On the vernal and autumnal equinoxes, approximately March 21 and September 21, all twenty-four temporal hours were sixty minutes long.

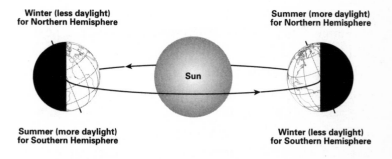

**Winter (less daylight)
for Northern Hemisphere**

**Summer (more daylight)
for Northern Hemisphere**

Sun

**Summer (more daylight)
for Southern Hemisphere**

**Winter (less daylight)
for Southern Hemisphere**

The earth's tilt causes the length of the daylight period to change over the course of the year.

The use of temporal hours, which varied in length depending on the day of the year, continued for many centuries. At the equator, the duration of day and night is equal every day of the year, but the farther a location is from the equator, the longer the days are in summer and the shorter they are in winter. So although the variation in the length of temporal hours may not have been terribly obvious to many Egyptians, it was more apparent to later cultures, as civilization spread north, away from the equator. During the Roman Empire, for instance, the summer solstice at the Coliseum consisted of twelve 76-minute daytime hours and twelve 44-minute nighttime hours.

ABUL-HASSAN AND THE ASTRONOMER ROYAL

Although the ancient Greek astronomer Hipparchus had, for astronomical purposes, divided the day into twenty-four equal *equinoctial hours* (based on the duration of temporal hours that occurs on each equinox), unequal temporal hours continued to be used in Europe through most of the Middle Ages. As late as the thirteenth century, the length of an hour in medieval London still varied widely throughout the year, from thirty-eight minutes to eighty-two minutes, and thus a daytime hour in summer was more than twice as long as a daytime hour in winter.

During that century, however, an Arab scientist working in Cairo began a transformation in timekeeping. Abul-Hassan al-Marrakushi, who designed sundials based on complex trigonometric principles, is generally credited with introducing the idea of a day of twenty-four equal-length hours for common usage. Not until the fourteenth and fifteenth centuries, however, did Abul-Hassan's equal-length hours come into general use in Europe, with the spread of mechanical clocks.

The mechanical clock, invented around 1270, was much better suited to displaying hours of fixed equal intervals of time than hours that varied in length between day and night and from one day to the next. The use of mechanical clocks spread slowly. Public clocks first appeared in London in about 1292, in Paris in 1300, and in Padua, Italy, in 1344. Clocks high up on town hall and church towers eventually proclaimed the local time in most cities and towns throughout Europe and beyond.

The first instance where the time shown by a clock was officially advanced one hour is said to have occurred in Switzerland in the fifteenth century, although it had nothing to do with saving daylight. In 1444, during a siege of the Swiss city of Basel, the invading army hatched a plan to overthrow the defending garrison. A simultaneous attack by the outside army and internal conspirators would be launched when the great town clock struck twelve noon. The watchman in the tower learned of the plot too late to warn the garrison, but thinking quickly, he set the hand of the clock ahead one hour. Expecting to hear the clock strike twelve times, the conspirators were thrown into confusion when they heard the clock strike only once. The attack was repelled, the conspirators were routed, and Basel was saved. In commemoration of this event the citizens of Basel declared that the town clock should remain one hour ahead, and this tradition continued for over three hundred years, to the end of the eighteenth century.

Early mechanical clocks were not particularly accurate, and until the 1600s most clocks had only one hand and were correct to within a quarter hour at best. Gradually clockmakers developed better mechanisms, and precision improved. Large public clocks were set by observing local noon—the time of day when the sun reached its highest point above the horizon—and individuals set their own clocks the same way or synchronized them with the clock in the town square. Thus each town followed its own "sun time" (called *apparent solar time*) based on its own local noon.

As the eighteenth century closed, clocks and watches had become more common and more accurate. But as timepieces became increasingly precise, a problem surfaced that had previously been masked by their very imprecision: An accurate clock shows every day as exactly twenty-four hours, but the actual length of each day is *not* exactly the same.

Due to the eccentricity of the earth's orbit and the tilt of its axis, the time from one day's local noon to the following day's local noon can be somewhat more or less than twenty-four clock-hours, depending on the day of the year. For example, the time on an accurate clock can be ahead of the local sun time as shown on a sundial by as much as fourteen minutes in mid-February and can lag behind sun time by as much as sixteen minutes in early November. In fact, there are only four days of the year when the clock and the sun completely agree. The difference between sun time and clock time, called the *equation of time*, was originally calculated in about 1670 by John Flamsteed, Britain's first royal astronomer.

Given the regularity of the clock and the irregularity of the observed sun, a perfectly accurate clock would have to be reset each day at noon. To avoid this, cities and towns began to set their clocks on the basis of *mean time*: the length of a mean-time day is defined as the average length of all the days of the year. Mean time (or *mean solar time*) was the first artificial adjustment made to natural sun time.

GUNS, BELLS, AND TIME BALLS

Mean time was first instituted in Geneva, Switzerland in 1780, and eventually most cities and towns followed suit. Even after mean local time was generally adopted, however, there was still the problem of keeping the population of a large city or region synchronized. Although more accurate clocks and watches were produced, as the nineteenth century progressed, they still could drift several minutes a day. Mean local time could be determined with the greatest precision by astronomical observatories that tracked *clock stars*, stars that appeared overhead each night at predictable times. In an effort to keep clocks and watches accurate, observatory time was often announced by firing a gun or ringing a bell each day at a designated hour or by dropping a *time ball*.

Time balls were large metal spheres that were dropped each day from a prominent building or tower at a precise time, often twelve noon. The exact time was relayed by telegraph from a nearby observatory. Time balls were first used to signal a precise time to ships at harbor, so each ship could set its chronometer accurately without having to send someone ashore. The Royal Greenwich Observatory began dropping a daily time ball as early as 1833. Soon a time ball was in use in many cities, so that at the designated hour observers at numerous vantage points could set their clocks to the accurate local observatory time. Thus the daily drop of the time ball fostered a uniform time for everyone in the area. A vestige of this practice is the illuminated ball dropped in New York City's Times Square at exactly midnight each New Year's Eve.

The use of mean local sun time and devices such as time balls allowed residents of each town or city to be synchronized, but there still was no coordination of times between different cities and regions. To understand how such a system might be possible, we need to consider that the relative sun times of two places is determined by their location on the globe. The ancient Greek astronomer, Hipparchus, was the first to imagine superimposing a grid on the earth's surface; his grid consisted of 360 lines (corresponding to the degrees of a circle) connecting the North and South Poles at right angles to the equator, and 180 equally spaced lines circling the earth parallel to the equator. The lines running between the Poles indicated a location's *longitude*, and the lines parallel to the equator indicated its *latitude*. The lines of longitude were later called *meridians*, from the Latin *meridies* (midday), because all places on the same meridian had local noon, when the sun is at its highest point, at the same time.

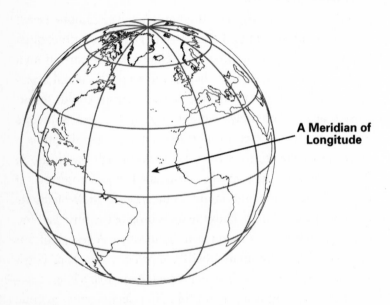

A Meridian of Longitude

A meridian is a line of longitude from pole to pole.

Latitude is measured north or south from the equator. For longitude, however, there is no obvious starting point. Therefore it is measured east or west from some designated line of longitude, and this is called the prime meridian. Up to the end of the nineteenth century, almost every major nation based its maps for land delineation and ship navigation upon its own defined prime meridian of longitude, usually the meridian through its capital city. Britain's prime meridian went through London, Portugal's through Lisbon, France's through Paris,

Russia's through St. Petersburg, and the United States' through Washington, D.C. To allow precise determination of longitude, the specific location of the prime meridian was usually located at an astronomical observatory in or near the capital city: the Royal Greenwich Observatory in Greenwich, England, just outside London; the Naval Observatory in Washington, D.C.; and the Pulkovo Observatory near St. Petersburg.

As the earth rotates, the sun appears to traverse fifteen degrees of longitude in one hour. Thus, each degree of longitude to the west, local noon occurs four minutes later. Consequently, any two cities not on the same meridian would have their clocks set at different times, depending on how many degrees their longitudes separate them.

Even though each town determined its time independently, the worldwide system of local times worked quite effectively for several centuries. As long as travel and communications were relatively slow, it didn't much matter that, for instance, in the United States when it was 12:00 noon in Chicago it was 12:31 in Pittsburgh, 12:24 in Cleveland, 12:17 in Toledo, 12:13 in Cincinnati, 12:09 in Louisville, 12:07 in Indianapolis, 11:50 in St. Louis, 11:48 in Dubuque, 11:39 in St. Paul, and 11:27 in Omaha. The relaxed pace of travel, the lack of instant communications, the inherent inaccuracy of contemporary clocks, and the less frantic pace of life all made minor time variations unimportant.

But then came the Industrial Revolution.

EIGHT MINUTES BEFORE CHIPPENHAM

Beginning in the mid-eighteenth century, the introduction of steam power and automated machinery ushered in a period of great technological progress. With it came sweeping social and economic shifts, including new forms of transportation such as the railroad (the locomotive was invented by Englishman George Stephenson, in 1814) and new means of communication such as the telegraph (developed by American Samuel Morse, in 1837). As the use of these revolutionary new technologies increased, the inexact nature of the long-established local time system became problematic. Time references in a cable dispatch could easily be misinterpreted. There was uncertainty as to whether a company or government office being cabled would be open for business when the cable arrived, and if a message was sent late, culpability could be concealed through the indeterminate differences in time.

It was the burgeoning railroad systems that experienced the most trouble. Prior to the construction of railroads, people had generally traveled infrequently and over short distances, and when they made an occasional long journey by horseback, coach, or ship, the distance traversed in a few hours or a day was small enough that the differences in local times did not present a substantial concern. Train travel, however, was much faster than any previous mode of transportation—fast enough that the time differences between nearby communities were noticeable. And since virtually every city and town used its own mean sun time, the patchwork of local times that was created made it very difficult for railroads to make understandable timetables, to coordinate shipments, or to schedule trains.

For a train traveling primarily north-south, local time differences didn't present a major problem, because the local times of towns at about the same longitude did not differ by more than a few minutes. But for trains going east-west, the local times differed significantly. This caused confusion for railroad personnel and travelers alike. Trains often kept the time of their original city of departure, but that frequently led to mix-ups with the various local times along the way. Timetables using local time were hard to understand and made east-to-west journeys seem faster than those from west to east.

To avoid these problems, toward the middle of the nineteenth century European railroads began to consider standardizing time by using the same clock time over an extended region. The idea had originally been proposed in the early nineteenth century by Dr. William Hyde Wollaston, a distinguished English physician and chemist (he discovered the elements palladium and rhodium). The financial security that resulted from his work in chemistry freed Wollaston to tackle other areas of scientific interest. His interest in time uniformity correlated with his involvement in attempts to bring uniformity to the systems of weights and measures and his active membership on the Board of Longitude, the government body formed to solve the problem of finding longitude at sea.

The cause of time standardization was then taken up in 1828 by Sir John Herschel, a well-known astronomer and early researcher in photography. In the 1830s the leading proponent of the concept was Captain Basil Hall of the Royal Navy, a famed explorer who had served as commissioner for longitude. In November 1840, England's Great Western Railway became the world's first rail system to adopt a single "railroad time": It instituted London Mean Time for all its operations throughout the country. In the next ten years or so, almost all other railroads in Britain did the same (although one English railway company initially refused to make public its new, single-time-zone timetables, anticipating that such action "would tend to make punctuality a sort of obligation").

The use of a single mean time over a large time zone was the second of the artificial adjustments to natural sun time. At first it was called *railroad time*, but eventually it became known as *standard time*. Standard time made scheduling and coordination much easier, as the British railroads soon found, but it moved timekeeping another step further from the natural observance of time.

Eventually railroads in several other countries, including Holland and France, followed the British example and established a single railroad time, usually based on the mean time at their country's prime meridian. In 1847, the Railway Clearing House, a British railroad standards organization, recommended that all railways in Britain follow Greenwich Mean Time (GMT); those that had originally followed London Mean Time now made the twenty-three-second adjustment to GMT. Dutch railroads used the mean time of Amsterdam, and French railroads adopted Rouen time, which was five minutes behind the time in Paris.

When the railroads in a country established a single time standard, several other institutions quickly adopted railroad time for their own purposes. Even so, local time continued in extensive use as well. Railroad passengers still had to account for local time as well as railroad time as they moved between trains and towns. Some watchmakers began making watches with two dials, one for local time and one for railroad time, and the great Tom Tower Clock, in Oxford, England, was fitted with two minute hands. In an effort to be comprehensive, one British railroad timetable in 1840 informed passengers, "London time is kept at all stations on the railroad, which is 4 minutes earlier than Reading time, 5½ minutes before Steventon time, 7½ minutes before Cirencester time, 8 minutes before Chippenham time, and 14 minutes before Bridgewater time."

In most cases there was no government statute making either railroad time or local time the official time—a situation that inevitably led to confusion. In 1859, a defendant in a lawsuit in

Dorchester, England, arrived at court at the appointed hour but found that the case had already been decided against him for nonappearance because the court used Greenwich Time while he had followed local time. The Court of Exchequer later ruled that local time was the legal time—"Ten o'clock is ten o'clock according to the time of the place"—and granted him a new trial. In France, railroad time continued to be five minutes behind Paris time, and travelers always had to keep in mind that the time inside a railroad station, *l'heure de la gare*, was five minutes behind the time just outside the station, *l'heure de la ville*. This situation continued for many years, even after Paris time was instituted as the national time for all of France.

A LINE OF CLOCKS STATIONED EVERY FIFTEEN DEGREES

A single countrywide railroad time could be established for relatively compact countries such as England, France, and Holland, where the local time at any location does not differ greatly from the time at the country's prime meridian. But in a very large country, such as the United States or Russia, there is a difference of several hours in local time between regions—between Boston and San Francisco or St. Petersburg and Vladivostok—and the use of a single national railroad time was not feasible.

In the United States, each railroad company adopted the local mean sun time of its home city or of some important cities on its line as its own standard time. The Pennsylvania Railroad used Philadelphia time. The Baltimore & Ohio operated on Baltimore time for trains originating in Baltimore and on Columbus time for Ohio trains. The New York Central operated in its Chicago district on Columbus time, which was nineteen minutes ahead of Chicago time, and the Union

Pacific Railroad had different time standards in six zones, according to the local time in Omaha, Jefferson City, St. Joseph, Denver, Laramie, and Salt Lake City. With rail lines proliferating, by 1872 there were more than seventy railroad "time zones" in the United States, with boundaries set according to each company's geographic coverage. Under such a system, it was not unusual for passengers traveling from Maine to California to adjust their watches more than twenty times en route, and in 1880, trains in Wisconsin alone were operating according to thirty-eight different railroad times.

This multiplicity of time standards confused passengers, shippers, and railroad personnel alike. "The confusion of time standards," grumbled the *New York Herald,* was "the source of unceasing annoyance and trouble." "The whole country was a pathless wilderness," *Harper's Weekly* later described the situa- tion. "Any traveler trying to wend his way across it was doomed to bewildering confusion. His watch was to him but a delu- sion; clocks in stations, staring each other in the face defiant of harmony both with one another and with the surrounding local time, and all wildly at variance with his watch, were wholly baffling to all intelligence, and time-tables were to him but Sphinx's riddles." Soon many railroad leaders were calling for some sort of national time standardization similar to the single railroad time that was in use in smaller countries.

Not only railroad-related interests, but also businessmen, telegraphers, and others encountered problems with the multi- plicity of times. So did the scientific community—especially astronomers and meteorologists, who were having trouble cal- culating, recording, and communicating information that had any relation to time because they had no standards to utilize. A number of proposals were put forward for a single uniform time for the entire United States, usually based on Washington,

D.C., or New York City time, but these did not get very far until, in 1869, Professor Charles F. Dowd, the principal of the Temple Grove Ladies Seminary in Saratoga Springs, New York, came up with a new idea.

Dr. Dowd was the first to conceive of a comprehensive, longitude-based standard time system for the United States. "England has only a difference of half an hour in meridian time," Dowd explained, "and France, computing from Paris, less than an hour. I could readily see how, in those countries, time could be governed from their prime meridians. But the great expanse of our country from the Atlantic to the Pacific would not admit of one arbitrary standard, for the difference is four hours of solar time." Then Dowd had a great insight: "It was while meditating upon this difficulty that the subject of hour sections first attracted my attention. As the sun describes the whole circle [of 360 degrees] in twenty-four hours, it occupies one hour in traversing fifteen degrees. Then, by starting from a certain standard meridian . . . the whole plan was clear to me. I could see that a line of clocks could be stationed every fifteen degrees across America, started with their minute hands at the same figure. Then, by placing the hour hand on the prime meridian at noon, that of each clock westward should be set back one hour for each fifteen degrees."

Dowd first presented this concept to a convention of railroad superintendents in New York City in October 1869. His "System of National Time" divided the country along straight meridian lines into four standard-time sections of fifteen degrees each:

> The Washington Section, for the Atlantic states, was centered on Washington, D.C., and used its local mean sun time.
> The First Hour Section, for the Mississippi Valley states, was one hour earlier than the Washington Section.

The Second Hour Section, for the Rocky Mountain states, was two hours earlier.

The Third Hour Section, for the Pacific states, was three hours earlier.

Charles Dowd's original plan for time zones.

With some encouragement—but no financial support—from the railroads, Dowd laboriously mapped the longitude of each of the eight thousand railroad stations in the United States and then calculated the difference between local time and the time in his corresponding standard-time section (see next page). During 1871 and 1872, he presented his findings to the New England Railway Association in Boston, the Western and Southern Railway Association in St. Louis, and other interested railroad organizations. Although Dowd's system was generally accepted by the railroad groups, questions arose as to the location of the standard meridians. For example, many Mississippi Valley railroad men felt that the standard meridian of Dowd's First Hour Section, which ran through Springfield, Missouri, was set too far west, which meant that sizable portions of their "western" railroads fell in the Washington Section. If the base of longitude were changed from Washington, D.C., to New York City, they suggested, the First Hour Section's standard meridian would fall halfway between St. Louis and Chicago and all the other time zone boundaries would shift east—to exactly where they wanted them.

ST. JOSEPH AND COUNCIL BLUFFS RAILWAY.

STATIONS.

[1]St. Joseph[+12]
[1]Amazonia[+12]
[1]Nodaway[+12]
[1]Forbes[+13]
[1]Forest City[+13]
[1]Bigelow[+13]
[1]Craig[+13]
[1]Corning[+14]
[1]Phelps[+14]
[1]Watson[+14]
[1]Hamburg[+15]
[1]E. Nebraska City[+15]
[1]Percival[+15]
[1]Bartlett[+15]
[1]Pacific[+15]
[1]Trader's Point[+15]
[1]Council Bluffs[+15]
[1]Omaha[+16]

Dowd compiled each station's proposed time zone and the time difference from local time. For example, "[1]Omaha[+16]" means Omaha's railroad time is one hour earlier than Washington time (i.e., in the First Hour Section), which is sixteen minutes later than Omaha's local time.

Dowd objected to using the New York meridian as a base because it was the local meridian of one American city and not the Washington, D. C., "national meridian," which had been established as the U.S. prime meridian years earlier. Instead, he supported another suggestion—a time zone system using longitudes based on the "nautical" meridian of Greenwich, England, which was by then becoming an unofficial international maritime standard. On many American maps, Greenwich-based longitudes were marked alongside Washington-based longitudes.

Dowd saw Greenwich-based time zones as a compromise, as it would move the time zone borders east two degrees, whereas a shift to the New York meridian would move them three degrees east. After some initial reluctance by the proponents of the New York–based system, the Greenwich-based proposal met with general approval and in the spring of 1872 Dowd modified his plan accordingly. His revised time zones were centered on and used the local mean times of longitudes 75°, 90°, 105°, and 120° west, based on the Greenwich prime meridian.

THE SURVEYOR'S ASSISTANT

Dowd spent years gathering support for his standard-time plan, traveling, lecturing, and writing numerous letters, pamphlets, and articles in support of his scheme. He found that although train travel and telegraph communication were both booming, the notion of time for most people remained fixed on the use of local time. It seemed senseless to set noon time according to abstract meridian lines that might be located far away rather than determining noon on the basis of the local view of the sun's position in the sky, as it always had been done. Though most railroad officials could see the usefulness of a standard time system, most citizens thought it impractical.

But in scientific circles, plans based on Dowd's concept of four U.S. time zones were gaining adherents. Dr. Cleveland Abbe, an astronomer and meteorologist with the U.S. Signal Service, was probably the foremost advocate for standardized time among American scientists. Abbe, the first official U.S. government weather forecaster, had seen the need for standardization most vividly when studying the phenomenon of aurora borealis, or northern lights, and trying to bring together data from numerous local observations, all of them giving the observer's local time. Abbe became a vocal champion of standard time and chairman of the American Meteorological Society's Standard Time Committee, and won the support of AMS president F. A. P. Barnard, who was the president of Columbia University. Sandford Fleming, the chief engineer of the Government Railways of Canada, worked to gather support for standard time within the American Society of Civil Engineers. But while some scientists, especially astronomers, pushed for national legislation on time standardization, most saw little hope because the scientific community had virtually no political influence. It was up to the railroads to act.

A man named William Allen would lead the way. Allen had worked for the railroads his entire life, starting at age sixteen as a surveyor's assistant. He was now editor of the industry publication *Travelers' Official Guide*, a compendium of railroad timetables and time zones, and secretary of the two major organizations that oversaw industry standards, the General Time Convention and the Southern Railway Time Convention. Having read reports and letters from Cleveland Abbe and other scientists strongly endorsing time standardization, Allen saw the great benefits of a time zone system like Dowd's and became an enthusiastic advocate of standard time.

Allen soon found, however, that a serious rift had arisen between advocates of standard time in the railroad companies and those in the scientific community. Whereas scientists favored Greenwich-based time zones, railroad people were worried that the Greenwich Time Zone boundaries would split up their territories and cause confusion. Setting practice above theory, Allen proposed a system of Greenwich-based time zones whose borders, instead of being rigidly drawn along straight meridians, were irregular and took into account the long-established territories of the railroad companies. Under Allen's plan, only a few railroads would have to make significant schedule adjustments.

Using this compromise approach, Allen continued to advocate for a national system of standard time. At the General Time Convention meeting held in St. Louis in April 1883, he presented two U.S. railroad maps side by side. One reflected the current system and all its intricate confusion ("the barbarism of the past," Allen called it), and the other demonstrated the relative simplicity of his plan for standard time ("the enlightenment we hope for in the future").

Allen's scheme defined four time zones in the United States, to be called Eastern, Central, Mountain, and Pacific Time. The standard meridians were the same as in Dowd's Greenwich-based plan—75°, 90°, 105°, and 120° west longitudes (roughly the longitudes of Philadelphia, St. Louis, Denver, and Reno). Each zone was centered on its standard meridian, and the time at any point in the time zone was defined to be the local mean time at the standard meridian. For the railroads in Canada, the convention established the same four time zones plus a fifth for Canada's far eastern provinces, called Inter-Colonial Time (now known as Atlantic Time), which was centered at 60° west longitude.

Allen received near-unanimous support from the General Time Convention, and a few days later his plan was endorsed by the Southern Railway Time Convention. Six months later, on October 11, 1883, railroad officials meeting at the General Time Convention at the Grand Pacific Hotel in Chicago overwhelmingly gave final approval to adopt standard time for railroads in the United States and Canada.

THE DAY OF TWO NOONS

The conversion of the nation's many timetables to standard time was set to take place at noon on Sunday, November 18, 1883. A Sunday was chosen because, as the day with the fewest

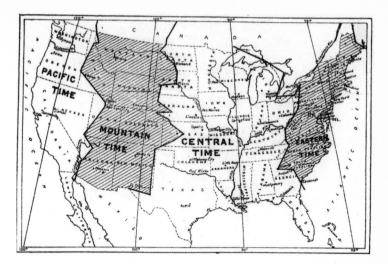

The original U.S. railroad time zones.

trains in operation, the transition would cause the least inconvenience and could be done most safely. William Allen had sent detailed instructions to the managers of all one hundred thousand miles of U.S. railroads on the arrangements necessary for carrying out the conversion to standard time.

On what would become known as "The Day of Two Noons," people across the country gathered to synchronize their watches on the basis of lowered time balls, striking town clocks, and chiming clocks in jewelery shop windows. As noon came and went in the eastern portion of each new time zone, the clocks were turned back one to thirty minutes, to the new standard time, and a second noon approached. People in the eastern part of a time zone were "living a little of their lives over again," quipped the *New York Herald,* while those on the western side of the meridian jumped, "some of them, half an hour into the future. The man who goes to church in New York today will hug himself with delight to find that the noon service has been

RAILROADS.

OLD COLONY RAILROAD.
CHANGE OF
Standard Time.

On **SUNDAY NOON, Nov, 18, '83,**
The Standard Time for the running of trains of this road
will be changed to conform to the "Eastern Standard
Time" then to be adopted by the Cambridge Observatory, and generally throughout New England, and which
is sixteen minutes slow of the present time.
The change will be made at the hour of the NEW
12 M. (12.16 P.M. present time).
　　　　　　J. R. KENDRICK, Gen. Manager.
Boston, Nov. 8, 1883.

A railroad's notice of the change to standard time.

curtailed to the extent of nearly four minutes, while every old maid on Beacon Hill, in Boston, will rejoice tonight to discover that she is younger by almost sixteen minutes."

Many greeted standard time with enthusiasm. *Harper's Weekly* proclaimed that, for the first time, "all the clocks on the continent struck together. The minute-hands of all were in harmony with each other, and with those of all travelers' watches. Time-tables everywhere became perfectly intelligible." The *New York Tribune* declared: "No reasonable person can doubt that the change will be to the general advantage. To have four time standards of railroad time throughout the country instead of fifty-three will be an immense saving of labor and worry." But despite the widely favorable reactions, there was also apprehension in some quarters. In New England, for example, where the clocks were set back fifteen or more minutes, manufacturers worried about increased fuel bills in winter as a result of additional darkness in the late afternoon. Some questioned the authority of the railroads to enact such a major change. Mayor Frederick Cummings of Bangor, Maine, initially refused to recognize railroad standard time on the grounds that it was unconstitutional. He even threatened to use the police to prevent church bells from ringing on the new time, but popular feeling ran against him and he soon relented. The *Indianapolis Daily Sentinel* also expressed its displeasure: "The Railroad Convention, recently in session, determined . . . to have the clocks and watches in the United States set, run and regulated to suit the convenience of their particular branch of business. It was a bold stroke. To regulate the time of the Empire Republic of the world is an undertaking of magnificent proportions. Railroad time is to be the time of the future. The sun is no longer to boss the job. People — 55,000,000 of people — must eat, sleep and work as well as travel by railroad time." The *Sentinel* then became quite impassioned in its opposition: "It is a revolution, a revolt, a rebellion, anarchy, chaos. The sun will be requested to rise and set by railroad time. . . . People will have to marry by railroad time, and die by railroad time. . . . Banks will open and close by railroad time. . . . And it is useless to ask, What are you going to do about it? We presume the sun, moon and stars will make an attempt to ignore the orders of the Railroad Convention, but they, too, will have to give in at last."

Seventy of the one hundred largest cities in America implemented the new time zones almost immediately, and within a few years most institutions had adopted standard time, as had almost all states, cities, and towns throughout the country, either officially, by statute or ordinance, or unofficially, by general consensus. Pockets of noncompliance remained for years, but by and large, standard time was now *the* American time. Moreover, although national standard time would not be the subject of any federal legislation for another thirty-five years, it was utilized unofficially by the federal government for most administrative purposes.

For his unique efforts in moving the nation to standard time, Charles Dowd annually received passes for travel on every major railway in the country. Ironically, years later, in 1904, after a lifelong association with the railroads, Dr. Dowd was killed at a grade crossing in Saratoga Springs—struck dead beneath the wheels of a locomotive.

A STANDARD OF TIME-RECKONING THROUGHOUT THE WHOLE WORLD

Similar efforts to standardize time took place around the world. Although railroad standard time remained unofficial in America, in some countries the system was legally adopted. In 1880, Greenwich Mean Time was adopted as the legal time in Great Britain, and Dublin Mean Time (twenty-five minutes behind Greenwich) the legal time in Ireland. In most countries, standard-time systems were based on the national prime meridian and little thought was given to international coordination.

In 1876, Sandford Fleming, who had played an active role in introducing standard time in the United States and Canada, envisioned a worldwide system of twenty-four one-hour standard time zones encircling the globe. The time zones—fifteen-degree-wide strips of longitude—would be centered on standard meridians every fifteen degrees, starting from an internationally established prime meridian. Over the next several years Fleming pushed his plan through correspondence, lectures, and papers at scientific conferences.

In 1884, President Chester A. Arthur invited countries from around the world to send delegates to the International Meridian Conference to be held in Washington D.C., "for the purpose of discussing, and, if possible, fixing upon a meridian proper to be employed as a common zero of longitude and standard of time-reckoning throughout the whole world." Since many countries used their own system of longitude based on their own prime meridians, there were numerous different prime meridians in use, including Amsterdam, Cádiz (Spain), Christiana (near Oslo, Norway), Copenhagen, Ferro (Canary Islands), Greenwich, Lisbon, Naples, Paris, Pulkovo (near St. Petersburg), Rio de Janeiro, Stockholm, and Washington, D.C. But the increasing volume and speed of international commerce and communication as well as the demands for uniformity from the worldwide scientific community led many to recognize the need for a single international standard. When forty-one delegates from twenty-five nations descended on Washington in October 1884, less than one year after the establishment of standard time in the United States, among the five American representatives in attendance were such staunch supporters of standard time as William Allen and Cleveland Abbe. Sandford Fleming was in attendance as part of the British delegation.

During the lengthy debate, the United States proposed as the international prime meridian the one passing through Greenwich, the home of the Royal Observatory and already a standard with most of the world's long-distance mariners. Some countries, however, especially France, were concerned that the international standard would be too British and argued for a more "neutral" prime meridian. One French delegate vowed, "France will never agree to emblazon on her charts 'degrees west or east of Greenwich.'" Sandford Fleming then suggested a prime meridian at "Greenwich plus 180 degrees," a mid–Pacific Ocean location halfway around the globe from Greenwich that was neutral but would provide the same time zones, and thus many of the same benefits, as Greenwich. But in the face of continued support for the Greenwich standard, Fleming himself pointed out that Greenwich-based longitudes were "used by seventy-two percent of the whole floating commerce of the world." Finally, by a vote of 22 to 1, with only San Domingo (now the Dominican Republic) voting against and

France and Brazil abstaining in protest, Greenwich was selected as the prime meridian for the entire world. All longitude would now be measured from a 0° meridian "passing through the centre of the transit instrument at the Observatory of Greenwich."

A Greenwich-based time-zone system was already in place in Great Britain, the United States, Canada, and Sweden, but soon other countries began adopting the Greenwich-based time zones: Japan in 1888; Belgium and South Africa in 1892; Germany and Italy in 1893; Denmark, Norway, Bulgaria, and Switzerland in 1894; and Australia and New Zealand in 1895. It would take many years for Greenwich Mean Time to be accepted around the world. Ireland stayed on Dublin Mean Time until 1916, more than three decades after the International Meridian Conference. France, long a holdout, finally succumbed in 1911, but they officially called their new time "Paris Mean Time, retarded by 9 minutes and 21 seconds," which allowed them to have Greenwich Mean Time without giving it a British name.

Thus the countries of the world synchronized their clocks. Order triumphed over chaos, and there emerged a clear and relatively simple global system that was generally accepted for commerce, communication, transportation, and science. But that system was soon to be challenged by supporters of William Willett's novel idea to capture more hours of daylight—the third artificial adjustment to natural sun time.

Chapter Three

The First Time

There is one thing stronger than all the armies in the world,
and that is an idea whose time has come.

—Victor Hugo

On July 28, 1914, World War I erupted in Europe. Austria-Hungary declared war on Serbia, holding Serbia responsible for the assassination, in Sarajevo, of Archduke Franz Ferdinand, heir to the Austro-Hungarian throne. Existing alliances, long-standing enmities, and competing interests would soon plunge other countries into the war one by one.

Within three days of the assassination, Russia, which shared a common Slavic heritage with the Serbs and opposed Austrian expansion, announced full mobilization of its armed forces. The next day, August 1, Austria-Hungary's principal ally, Germany, mobilized its armed forces and declared war on Russia. Two days later, Germany declared war on Russia's ally, France. Seeking a quick victory, German troops immediately marched into Belgium as a flanking maneuver to strike at France. Germany had hoped that Britain would remain neutral, but the German invasion of Belgium, a British ally, brought a British declaration of war the very next day, August 4.

In little more than a week, the major European countries

were at war with each other—Britain, France, and Russia on one side ("the Allies"), and Germany and Austria-Hungary on the other ("the Central Powers"). Over time, the war would stretch around the world to entangle numerous nations and become the largest conflict the world had seen to that time, involving tens of millions of soldiers and affecting untold hundreds of millions of people.

The Great War changed everything. "In the crush of that war," Winston Churchill would later observe, "people were forced to give up old prejudices and shut off the sluggish inertia of their minds." New issues rose to importance, conventional ideas could be challenged, and untested concepts could be reevaluated.

BRITAIN BIDES, GERMANY DECIDES

The death of William Willett in 1915 seemed to end any serious consideration of daylight saving time, but with the advent of war, the view of DST changed significantly. While adding hours of evening recreation was naturally of less importance now in the warring countries, there was increased concern about the conservation of energy for the war effort. Fuel, primarily coal, needed for generating electric current was scarce, and the use of artificial lighting had to be cut back wherever possible. The adoption of daylight saving time would allow factories, businesses, and households to take advantage of sunlight for an additional hour each day.

A year and a half into the war, on February 16, 1916, Sir Basil Peto, member of Parliament for the town of Devizes in Wiltshire, questioned Prime Minister Herbert Asquith in Parliament regarding the need to conserve electricity, gas, and oil for the war effort. Would the prime minister, Sir Basil pressed, propose "legislation on the lines of the late Mr. Willett's Daylight

Saving Bill?" "No, Sir," Asquith replied dismissively, "I cannot introduce legislation on this contentious subject." A brief three weeks later, on March 7, the prime minister fielded more questions about DST, including a suggestion that a committee of experts study the issue. Citing recent wartime measures, Prime Minister Asquith once again rebuffed the suggestion, declaring, "The joint effect of the darkening of the streets and the early closing of places where intoxicants are sold has probably contributed more towards shortening the interval between sunset and bedtime than would the adoption of Central European Time as the standard time during the summer. I do not think that there is sufficient reason to appoint a Committee of Inquiry on this subject."

While the British debated, however, Kaiser Wilhelm II's Germany took action. On April 6, 1916, the German Federal Council, operating under its emergency powers, decreed that summer daylight saving time would be instituted in Germany as a wartime measure, starting the last Sunday of that month. German *Sommerzeit* ("summer time") was to begin at 11:00 P.M. on Sunday, April 30, 1916, when clocks would be advanced one hour from Germany's Central European Time. *Sommerzeit* would remain in effect until 1:00 A.M. on Sunday, October 1, at which point clocks would be set back one hour. Concerning this forward-looking action, the *Frankfurter Zeitung* newspaper pointed out that, although there had been an English daylight saving time movement for many years, DST had never been employed there. "It is characteristic of England," the newspaper's editors observed, "that she could not rouse herself to a decision."

German business and industry anticipated the adjustment, and advertisers were quick to point out the material benefits of the time change. One newspaper ad announced, "One hour earlier than usual, you can obtain on May 1 our beautiful and much-coveted summer clothes. One hour earlier than usual, you can acquire the happy feeling of having bought from us, and therefore to your advantage. One hour earlier than usual you can go out into the May sunshine in full enjoyment of your pretty new dress." Not missing a beat, on April 30, the evening when, at 11:00 P.M. the clocks were to be set forward to midnight, the Berlin Opera began its performance of *Die Meistersinger* an hour earlier than usual so that audience members might catch their customary trains.

Though no serious opposition was voiced, the German government took preemptive action and issued a strong warning to enterprises that might be tempted to nullify the new law by moving their opening and closing times forward an hour, thereby maintaining their previous schedules. Energetic measures would be taken against any such effort to evade the law while ostensibly observing it. Summing up the matter, the *Frankfurter Zeitung* enthusiastically noted the many benefits *Sommerzeit* bestowed on Germany: "The objects of summer time are in the first place of an economic kind. They lengthen the amount of daylight by about 150 hours, they spare light, and they make more intense labor possible. They are also of a hygienic and social kind, because people will get more of the beneficent rays of the sun, and the light evenings will make it easier for them to seek health in the open air. If the change involves certain inconveniences, they will be far outweighed by the great advantages. Summer time is indeed a precious gift. It is evidence of the new spirit that has arisen with the war—a spirit of community that overcomes all obstacles in order to achieve results of benefit to everybody."

Thus not Britain but Britain's greatest enemy at that time was the first to adopt a plan for daylight saving time. Under the pressure of war, neighboring countries in close political or commercial touch with Germany quickly followed its lead. Austria-Hungary, Germany's chief ally, established DST for a period coinciding precisely with Germany's. The commercial ties between Germany and neutral Holland led the Dutch to institute DST as well. In Denmark and Sweden advanced time went into effect two weeks after Germany's initiative.

WILLETT TIME

In Britain, with the Great War intensifying and in light of the recent adoption of daylight saving time by Germany and other European countries, there was continuing pressure for DST in Parliament and elsewhere. The mounting debate culminated in the House of Commons on May 8, 1916, when Sir Henry Norman, a DST supporter since the original Daylight Saving Bill of 1908, made a motion asking the government to introduce a DST bill. Sir Henry listed the many benefits of DST and finally brought up its relevance to the war: "Unhappily our enemies have been quicker than we have been to secure the great economy that resulted from the advancement of clock time. The measure has been in operation in Germany, Austria and Hungary from May 1." Concluding his remarks with a rhetorical flourish, he pointed out how regrettable it would be if "a measure born in this country should profit only our enemies." Sir Henry's motion carried 170 to 2, and the next day, only two months after the prime minister had refused even to consider a daylight saving time measure, Home Secretary Herbert Samuel, representing the government leadership, introduced a DST bill in Parliament.

To make his case, Samuel assured the House of Commons that the government would not have dreamed of favoring the proposal if they had not thought it essentially advantageous for the wartime economy. Because large numbers of miners had enlisted in the military, the output of coal was declining, and a move to DST would reduce consumption of fuel for artificial lighting. Likewise, Sir Owen Philipps, MP for Chester, contended that DST would help boost the nation's shipping capacity, since work on docks had to stop at dark, due to the threat of German Zeppelin raids. Home Secretary Samuel also observed that although the minds of all ought to be concentrated on the war, and not on amusement and opportunities for recreation, still, "healthfulness and cheerfulness did have their value in maintaining the morale of the nation."

Debate in Parliament focused on remaining in step with other European countries. "It would be an unfortunate but by no means unprecedented thing," Home Secretary Samuel observed, echoing Sir Henry Norman's earlier remark, "if the United Kingdom, which was the first to originate the idea, should, owing to its own slowness to move, be the last country in Western Europe to adopt it."

A few voices in the Commons were raised in opposition, and Lord Balfour of Burleigh proclaimed the Daylight Saving Bill the most ridiculous and absurd measure ever presented to the House of Lords. He highlighted some of its unique disadvantages. For example, he asked their lordships to consider the night in October when the clock was to be set back an hour, say from one o'clock to midnight. "Supposing some unfortunate lady was confined with twins, and one child was born ten minutes before one o'clock; if the clock was put back, the registration of the time of birth of the two children would be reversed." The elder child would be properly registered as being born at 12:50, but the younger child, born ten minutes later, would be registered as being born at 12:00. Lord Balfour argued, "Such an alteration might conceivably affect the property and titles in that house," which in England generally passed to the eldest son.

But the naysayers were in a distinct minority. Although lengthy debate had accompanied previously proposed DST legislation, on this occasion it took less than a week for Parliament to conclude its deliberations. It approved the bill on May 15 and royal assent made it a law on May 17, 1916 — eight years after Robert Pearce had introduced the first DST bill embodying Willett's ideas. The Summer Time Act of 1916 established British summer time — called "Willett Time" by some — for the year 1916 and authorized extensions for the duration of the war.

At the close of his presentation to Parliament on the bill, Home Secretary Samuel had reflected, "How often has it happened that a great architect died before the building he had designed was completed, and a great musician passed away before his *chef-d'oeuvre* had been performed? Among such incidents would rank the pathos of the death of William Willett, after so many years of labour in the cause, just at the moment before the adoption of his scheme over a large part of Europe and in the country in which he originated it."

Then Sir Henry Norman, a champion of Willett's plan from the beginning, added a few words of his own: "Nobody can speak of daylight saving on the eve, as I hope it may be, of its triumph, without a word of tribute to the memory of the late Mr. William Willett. It was he who originally conceived the idea. He gave years of his life and munificently of his means, to the advocacy of it. . . . No criticism daunted him. No defeat dismayed him. Apart from his own business, he lived for daylight saving. I have never known a more persistent, a more genial, a more disinterested advocate." Sir Henry concluded, "I venture to think that the time will come when the workers of this country will desire to erect to William Willett, in grateful memory, a statue on some peak, where it would be gilded by the first ray of an April sunrise — one hour in advance of Greenwich Mean Time." The House of Commons cheered heartily.

A BIG BEN-EFIT.

MR. ASQUITH—" Who said we don't economize? "

Prime Minister Asquith's government finally passed a daylight saving time bill.

THE SUMMER TIME ACT OF 1916
LOCAL TIME IN SUMMER MONTHS

1.—(1) During the prescribed period in each year in which this Act is in force the local time in Great Britain shall be one hour in advance of Greenwich Mean Time.

(2) This Act shall be in force in the year nineteen hundred and sixteen, and in that year the prescribed period shall be from two o'clock in the morning, Greenwich Mean Time, on Sunday the twenty-first day of May, until two o'clock in the morning, Greenwich Mean Time, on Sunday the first day of October, and His Majesty may in any subsequent year, by Order in Council made during the continuance of the present war, declare this Act to be in force during that year, and in such case the prescribed period in that year shall be such period as may be fixed by the Order in Council.

(3) Wherever any expression of time occurs in Act of Parliament, Order in Council, order, regulation, rule, or by-law, or in any deed, time-table, notice, advertisement, or other document, the time mentioned or referred to shall be held during the prescribed period to the time as fixed by this Act:
    Provided that where in consequence of this Act it is expedient that any time fixed by any by-law, regulation, or other instrument should be adjusted and such adjustment cannot be effected except after the lapse of a certain interval or in compliance with certain conditions, the appropriate Government Department may, on the application of the body or person by whom the by-laws, regulation, or other instrument was made or is administered, make such adjustment in the time so fixed as in the circumstances may seem to the Department proper, and if any question arises as to what Government Department is the appropriate Government Department the question shall be finally determined by the Treasury.

(4) This Act shall apply to Ireland in like manner as it applies to Great Britain, with the substitution, however, of reference to Dublin Mean Time for reference to Greenwich Mean Time.

(5) Nothing is this Act shall affect the use of Greenwich Mean Time for purposes of astronomy or navigation, or affect the construction of any document mentioning or referring to time in connexion with such purposes as aforesaid.

Summer time for the United Kingdom.

PROMENADE CONCERTS IN MAY

Under the Summer Time Act, British clocks were first advanced one hour at 2 A.M. on Sunday, May 21, 1916, only four days after final approval of the act and just three weeks after Germany's DST period had begun. *The Times* of London reported that children begged to be allowed to stay up on the preceding Saturday night to watch their fathers change the time on the clock. One imaginative parent delegated the time-changing duty to the youngest in the family, so that the boy would be able to tell another generation that he could recall the first ceremony of its kind.

Many public clocks had been set an hour ahead on the previous afternoon, which gave some afternoon strollers a start and caused a sudden dash to pubs, where patrons were relieved to find that there still was ample time in which to quench their thirst. According to one newspaper account, when Big Ben, the clock at Westminster, was advanced, passersby witnessed, "the unusual sight of the minute hands, each fourteen feet long, traveling rapidly around the dial followed by the hour hands, each nine feet long, at a more deliberate pace." The few long-distance trains traveling overnight ran one hour behind their timetables.

May 21, the first day of daylight saving time in the United Kingdom, dawned beautiful and sunny in London. The sunshine lasted all day and into the evening, prompting one enthusiast to declare, "It's not the clock but the calendar that has been moved up. Are you sure we haven't stepped right into June?"

Nonetheless, there were scattered expressions of disapproval of the time change. On the Saturday prior to the switch, a large meeting of farmers in the town of Northampton unanimously approved a resolution "to adhere as far as possible to real time as shown by the sun . . . and to take as little notice as we can to the sham time that will be shown by public clock."

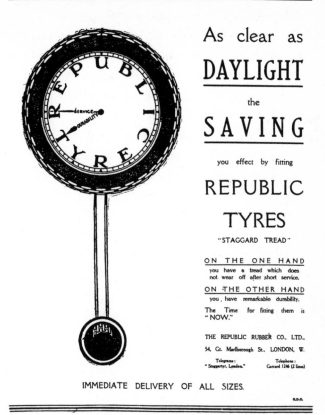

As clear as

DAYLIGHT

the

SAVING

you effect by fitting

REPUBLIC

TYRES

"STAGGARD TREAD"

ON THE ONE HAND
you have a tread which does
not wear off after short service.

ON THE OTHER HAND
you have remarkable durability.

The Time for fitting them is
" NOW."

THE REPUBLIC RUBBER CO., LTD.,

54, Gt. Marlborough St., LONDON, W.

Telegrams :   Telephone :
" Staggertyr, London."   Gerrard 1246 (2 lines)

IMMEDIATE DELIVERY OF ALL SIZES.

On Britain's first day of DST, advertisers woke up early.

Many munitions workers in Sheffield and other cities with
work hours starting at 6 A.M. overslept and were distressed to
find that they had lost some wages as a result. Night-shift workers
at the Devonport Dockyard thought they should be paid for the
hour of work time they had lost when the clock was advanced,
but they were unable to convince their superiors that the hour
from two to three o'clock had been worked. There were even
reports of an increase in gasoline consumption, since some

Britons utilized the additional hour of daylight to take a drive in their motorcars. And in a particularly stinging rebuke to patriotic feeling, a scribbler with the initials G.R. noted that the time shift had put Britain on its enemy's standard time and chalked on London's New Bridge Street: "All Fools' Day, May 21. Get up one hour earlier, and kid yourself you haven't Berlin time."

Beyond these few negatives, most reports of the first DST period from all over the country were supportive. Birmingham factory workers expressed hope that DST would be put into effect every year. Brighton's Promenade Concerts, which had been considered unfeasible until June because of dark evenings and stringent wartime lighting restrictions, could now commence. The city of Bradford estimated savings of £200 on gas bills during the first week of DST, and Nottingham reported a 10 percent reduction in gas consumption. Public bowling greens and tennis courts in Sheffield were kept open an hour later and were busy until closing. After only a week of operation, *The Times* declared, "Summer Time is already an accepted institution."

Just after the United Kingdom launched summer time in 1916, other European countries such as Norway, Italy, and Portugal also initiated DST. Near the end of 1916, even Australia decided to introduce daylight saving time during its Southern Hemisphere spring and summer months. In France, scientist Charles Lallemand presented to the Paris Academy of Sciences a critical report strongly opposing DST, but at a meeting of the Astronomical Society of France, astronomer Camille Flammarion laid out the practical merits of the proposal. Ultimately, after most other major nations in Europe had adopted daylight saving time, France decided to do so as well. *L'heure d'été* ("summer time") began on June 14,

1916–132 years after Franklin had recommended his "Economical Project" to Parisians.

A SUMMER IN THE SUN

The 1916 DST period in the United Kingdom lasted until 3 A.M., Sunday, October 1, when the clocks were set back to 2 A.M. Greenwich Mean Time. The previous May there had been no trouble setting the hands of clocks ahead one hour, but since the act of turning the hands backward could damage a clock's delicate chiming mechanism, it was preferable to move the hands ahead eleven hours, pausing each time the clock struck. This could be a long and laborious process, especially for a clock that struck every quarter hour. A few months later, a letter published in *Nature* proposed solving this problem by leaving the hands of the clock untouched but, instead, rotating the dial behind them backward in spring and forward in autumn—an innovation that never became popular with clockmakers or the public.

As Britain's first experience with daylight saving time came to an end, after years of debate and forecasts of great benefits and dire consequences, assessments of the actual impact of DST could finally be made.

In September 1916, Home Secretary Samuel set up the Summer Time Committee, chaired by MP John Wilson, to "enquire into the social and economic results of the Summer Time Act" and to recommend whether and how summer time should be continued in subsequent years. On the basis of a series of hearings and information gathered from more than a thousand surveys returned from individuals, organizations, and government authorities, the committee evaluated the impact of DST in a variety of areas.

A central goal had been decreasing fuel consumption, and

SUMMER TIME ENDING.

ADVICE ON PUTTING BACK THE CLOCK.

Ten days hence "summer time," in the horological sense, will come to an end. The Act under which the nation approximated its hours of work and recreation more closely to those of daylight by putting on the clock an hour at 2 in the morning on May 21 directs that the clock shall be put back an hour at 2 in the morning (Greenwich mean time : 3 o'clock, summer time) on October 1.

Posters calling attention to the resumption of normal time have been prepared by the Home Office and will be displayed on public buildings, at the railway stations, and elsewhere early next week. It is expected that the change will be effected with as little disturbance of the routine of life as the transition from normal time to "summer time" four months ago. The only difficulty at present foreseen is a technical one. Watches and non-striking clocks may be freely altered : but the greatest care will be needed in the case of striking clocks. An expert in these matters stated yesterday that on no account should the hands of such clocks be turned back when the time for the alteration comes, or serious damage may be done to the timepiece. The change can be made in one of two ways. Either the clock may be stopped for an hour, between two striking points, or the minute hand may be moved, with pauses at the striking points, 11 times round the dial.

This danger will be pointed out in the notice to be issued by the Home Office, but for all that there will probably be a considerable number of "casualties" among clocks on October 1, and the repairers expect to be kept busy for some time.

Setting the clock back was the difficult part of summer time.

gas and electric lighting companies were able to report a substantial decline in the use of artificial light, although an accurate estimate was difficult to make because of the exceptional wartime conditions, such as lighting restrictions and greatly increased overtime work. Overall, reductions in power for electric lighting averaged about 20 percent, which corresponded to a cutback of about 1 percent of the total coal used for all purposes in a year.

Many mothers noted the difficulty they experienced getting their children to bed when it was still light, but in general they believed the advantages of summer time more than outweighed the drawbacks. Schools reported a small number of children who appeared to be tired, but again, a large majority of school officials felt that this drawback was outweighed by the extra time children could spend in the open air.

Regarding DST's effect on public health, no definite conclusions could be reached under the abnormal conditions of a war. Nonetheless, it was widely assumed that the populace used the extra hour of daylight for outdoor recreation. A report from Bourneville stated, "Cricketers, bowlers, and tennis players took full advantage. . . . The Open Air Swimming Bath has been made much greater use of . . . and gardening classes for boys and girls [continued] a month longer than in previous years." The medical consensus was that the additional exercise must have proved beneficial.

Moreover, police authorities reported a definite improvement in public order. Policing is always easier during daylight hours, and a marked decrease was found in juvenile crime, which the police attributed to it still being light at the time when boys were expected home. Furthermore, there was a definite (though slight) decrease in street accidents.

Employers in most businesses throughout the country overwhelmingly favored summer time, as did the majority of workers and trade unions, though workers in certain industries complained of hardship. In the cotton-weaving sheds in Lancashire, for example, the need for artificial light in the early morning, especially in late September, caused such high temperatures in the sheds that workers refused to work in them. It had been thought that shopkeepers might keep their shops open an extra hour under summer time, thus depriving sales clerks of any benefit, but in fact reports indicated that only about 5 percent of shops extended their hours, and it was felt that this practice was unlikely to spread beyond a very small and not overly reputable minority of shopkeepers.

Opinion about DST was especially strong among farmers. Most often mentioned was the difficulty of harvesting before the heavy morning dews had lifted. Farmhands had to start milking early in order to meet the summer time schedules of milk delivery trains, but then they had to wait for the fields to dry before they could start harvesting corn or hay. On the other hand, fruit and vegetable growers reported approvingly that gathering and transporting produce was easier during the cooler morning.

In certain agricultural districts, farms did not observe the Summer Time Act at all, while local institutions such as schools and post offices did observe it, which naturally caused confusion and friction. Overall, however, in spite of the difficulties it caused, a large majority of farmers and agricultural committees were in favor of renewing the Summer Time Act, and most of those who opposed it still recognized the benefits of DST to the community at large.

Sir William Napier Shaw, director of the Meteorological Service, reported to the committee that "in spite of very careful instructions, a great deal of confusion arose as regards the hours at which [meteorological] observations were made." Meteorological records continued to be collected by telegraph from one hundred official and five hundred volunteer sites using Greenwich Mean Time, not summer time, as specified by the Summer Time Act. But records from many sites were telegraphed just before the closing of local post offices, and now that post offices closed an hour earlier by Greenwich Time, the continuity of weather records over the years was spoiled.

Although many had anticipated confusion arising when the actual time changes were made, the committee found that the transitions from Greenwich Time to summer time and back

PODGERS—(as alarm goes) "Liar!"

Summer time caused some inconvenience.

were relatively seamless. The only real problem came from people forcibly setting back their clocks, in spite of official warnings on the subject, leading to "a number of casualties among striking clocks."

After considering all the evidence it had gathered, the Summer Time Committee concluded that it could "unhesitatingly say that the vast preponderance of opinion throughout Great Britain is enthusiastically in favor of summer time and its renewal, not only as a war measure, but as a permanent institution." The committee issued its final recommendations in February 1917:

1. That summer time should be renewed in 1917 and in subsequent years.
2. That the period of operation of summer time should be from the second Sunday in April to the third Sunday in September.
3. That the change from normal to summer time should be made on the night of Saturday-Sunday and the revision to normal time on the night of Sunday-Monday.
4. That the variation from normal time should remain one hour throughout the whole period.

Carrying out the committee's recommendations, in October 1917 Parliament established summer time each year for the remainder of the war.

SOMMERZEIT AND L'HEURE D'ÉTÉ

Like officials in Britain, their counterparts on the Continent evaluated summer time. The German government and a number of German and Austrian newspapers made detailed inquiries regarding the impact of their experiments with daylight saving time. Their findings from commercial and industrial sources, from city and town governments, and from the general public were supportive of *Sommerzeit*. "Positive views have been expressed as to the value of extending the afternoon hours and thereby providing for the recreation and the benefit of being in the fresh air," the Berlin Chamber of Commerce reported. It recommended "most earnestly the retention of the summer period" and asked that "instead of May, the period should be fixed in the early part of April."

Several large cities reported significant savings of artificial light and fuel; in Berlin, gas usage decreased by half a million cubic meters for the *Sommerzeit* months of May and June compared to the previous year. As in England, some negative effects were reported in the rural districts of Germany: Agricultural interests strongly opposed *Sommerzeit* and schoolchildren were found to have suffered from lost sleep.

"Daylight saving was a great success in Austria-Hungary last summer," declared Albert Halstead, the American consul general in Vienna, "proving most beneficial to the health of the residents of Vienna because of an extra hour of sunlight in working hours, which did much to save lighting expense." He reported that five months of daylight saving in 1916 lowered gas consumption by the equivalent of $2 million in today's dollars in Vienna alone.

In France, the first summer of *l'heure d'été* was also considered a success. Chambers of commerce and labor organizations throughout the nation were enthusiastic, although there was the typical opposition by farmers. French shipping authorities reported that the loading capacity of the docks at Calais increased by 250 tons a day because longshoremen were willing to work overtime for the extra hour. Paris, Toulouse, Nîmes, Dijon, and Blois all recorded less use of electricity and together were thought to have saved more than 400,000 tons of coal, worth over $125 million today.

On the basis of the first summer's results, Germany, Austria-Hungary, France, Holland, Italy, Portugal, and several other countries continued daylight saving time each summer through the end of the war. Scandinavia already enjoyed long, bright summer evenings without having to advance the clocks, so fuel savings had been insignificant, and Norway, Sweden, and Denmark decided to discontinue DST. Australia subsequently joined the Scandinavian countries by repealing DST countrywide, but Spain and Russia, neither of which had tried DST in 1916, adopted it the following year.

While many European nations, through some combination of argument and necessity, had become convinced of the benefits of turning forward the clock, one conspicuous country across the Atlantic had failed to act. When Willett and Pearce first introduced the daylight saving bill to Parliament in 1908, *Scientific American* predicted "Whether the English measure be passed or not, it is unlikely that any daylight bill of this kind will be introduced into the United States, as least for many decades to come. Tradition, habit, and a hundred settled usages, national, commercial, and domestic, will always be ready with a strong protest against any interference with that symbol of unchanging order, the clock."

But the raging world war was soon to engulf the United States as well, and with war would come increased interest in daylight saving time and its potential benefits.

Chapter Four

America Takes Its Time

In war, time is vital.

—British Prime Minister David Lloyd George,

"Message to America," March 1918

*I*n 1916, as the war in Europe continued, bloody conflicts such as the Battle of Verdun lasted for months and claimed upward of one million casualties. But as the armies clashed, America remained at peace. President Woodrow Wilson warned Germany to discontinue its unrestricted submarine warfare policies, whereby neutral American vessels were targeted, but then ran for and won reelection with the campaign slogan "He kept us out of war."

As the war combatants began instituting daylight saving time, Americans paid little attention. The war was still an ocean away, and although standard time was now in use throughout the country, there had been no national time legislation of any kind. Nonetheless, news of William Willett's idea had gradually drifted across the Atlantic, and in the United States a modest but growing movement for daylight saving time had begun.

In 1909, Willett had written a letter to every member of

Congress pressing for the adoption of daylight saving time in America. His motive was to quash an argument being raised against his DST proposal in England. Some believed that if the United Kingdom enacted daylight saving time and the United States did not, it would have a detrimental effect on Britain's financial institutions because the London and Liverpool Stock Exchanges would cease to share an hour of simultaneous operation with the New York Stock Exchange. But Willett's letter was unpersuasive. "We cannot see why if any great number of people want to begin and stop work earlier on summer days, they shouldn't go ahead and do it, without any childish fooling," editorialized the *New York Times*. Willett's proposal, the paper concluded, was "little less than an act of madness."

But a few Americans fastened upon the idea. E. H. Murdock, a prominent businessman in Cincinnati, Ohio, and president of the Queen City Printing Ink Company, was active in civic and charitable affairs. Murdock had learned of Willett's concept on a trip to England and grown quite excited about it. Upon his return, he enthusiastically spread the word about daylight saving time and was able to arouse considerable public interest among his fellow Cincinnatians. Under Murdock's leadership, local DST adherents banded together to form the National Daylight Association of Cincinnati, the first daylight saving time advocacy group in the United States. The group's goal was "the saving of one hour of daylight each day for the five summer months, May 1 to October 1, of each year to all the people of the United States."

In pursuit of their goal, Murdock, J. G. Schmidlapp, the president of the Union Savings Bank, and six other members of the Cincinnati group arranged a meeting with President William Howard Taft to argue their case. President Taft—large, jovial, and conscientious—received the group at the White

House on May 17, 1909, and talked with them at length. But though the president was intrigued by the DST concept, he told the group that he could not propose national legislation at that time and encouraged them, as a first step, to pursue their daylight saving plan at a local level in Cincinnati. Taft even went so far as to propose at his cabinet meeting the next day the possibility of shifting government employees' working hours an hour earlier to give more daylight in the workday, though his cabinet ultimately concluded that the promoters of daylight saving would have to take the matter up with Congress.

In fact, a week after the Cincinnati group met with the president, Representative Andrew Peters of Massachusetts introduced a bill in Congress proposing nationwide DST. But with little interest or support for DST across the country, the legislation received scant notice and quickly died in committee.

President Taft's encouragement prompted Murdock and the National Daylight Association to attempt to garner support for local DST in Cincinnati. They also launched an effort to form daylight associations in all sections of the country to promote local DST, with the goal of growing a national daylight saving movement and eventually enacting a DST measure for the entire United States. These endeavors met with little tangible success in Cincinnati or anywhere else, but they did help introduce the concept around the country. Some would later call Murdock the father of the daylight saving movement in America.

During the next few years, opposition was strong whenever new proposals for DST surfaced. William Allen, who thirty years earlier had led the campaign for railroad standard time in America, was firmly opposed to the idea. "It is futile to suppose that anything will be gained by such a change in the hands of the clock," he asserted. "The idea that there will be gain is like that of a man cheating himself at solitaire and thinking he has won the game." Whereas daylight saving time had become an issue of national controversy in Britain, in America it remained little more than a curiosity, receiving little notice and exciting limited interest.

A TALE OF TWO CITIES

As the concept of advancing clock time slowly spread around the country, the cities of Detroit and Cleveland came to play a significant role in the push for "more daylight," though not exactly in the manner that the DST advocates proposed.

Both of these cities were situated in the Central time zone, both at longitudes about halfway between the standard meridians of the Eastern and Central time zones—Cleveland's sun time was 27 minutes behind Eastern Time and 33 minutes ahead of Central Time, and Detroit's was 32 minutes behind Eastern and 28 ahead of Central. Their positions in the Central time zone kept them synchronized with Chicago and the local railroads. But in the view of some locals, a permanent shift to eastern standard time would benefit the cities by providing more daylight for evening recreation while putting them on the same time as New York and other major eastern financial centers. By changing from Central Time to the Eastern Time and permanently moving their clocks ahead one hour, both cities would essentially give themselves year-round daylight saving time.

This movement for a time zone change had begun modestly in Detroit in 1907, with the organization of the More-Daylight Club. The group's goal was to promote shifting Detroit to Eastern Time to increase usable daylight. Its initial meeting was attended by just two people: George Renaud, a well-known physician who specialized in hay fever and throat ailments, and C. M. Hayes, Detroit's leading portrait photographer. This dedicated duo gradually attracted more time-shift adherents, yet "owing to general ignorance and apathy, and lack of newspaper support," Dr. Renaud reported, "public interest was very slow to be awakened." They initiated their campaign without knowledge of William Willett's efforts in England, but before long Renaud and Hayes would embrace many of Willett's arguments and would even borrow his title, *The Waste of Daylight*, for their own publication.

Their movement persevered, Renaud noted, in the face of "the opposition of the press, the Board of Commerce, organized labor, and every organization approached," and in the fall of 1908 the More-Daylight Club led an effort to force a city-wide vote on a change to Eastern Time. The proposal was defeated soundly, losing in every one of the 150 voting precincts in Detroit's eighteen wards. Undeterred, the More-Daylighters continued their campaign and after three years resubmitted the issue to a vote. In 1911 the proposal was again rejected, but this time it carried eight of the eighteen wards.

In a manufacturing city such as Detroit, where people were already somewhat removed from the rural rhythms of nature, the More-Daylighters felt that a time zone change would bring more usable daylight because people would follow the clock, not the sun: "We get up in the morning, we go to bed at night, we go to work and quit, and we eat, when the clock says to." One of the group's goals was to plant the seeds of change in other similarly situated Central time zone cities; their hope was ultimately to extend the Eastern time zone by moving its boundary west. The group was persistent in its efforts, and "in Cleveland," Renaud observed, "the seed fell on good soil."

There a similar movement blossomed. The idea of shifting Cleveland to Eastern Time was aggressively pursued by the local chamber of commerce, was backed by the local press and labor unions, and was endorsed by the city's amateur football and amateur baseball associations, who appreciated the prospect of more light in the late afternoon. Proponents pointed out that Cleveland's earliest sunrise, 3:51 A.M., was too early to benefit many Clevelanders, and its earliest sunset, 3:55 P.M., deprived most of sunlight after working hours. If one assumed that the average hours of rising and retiring were 6 A.M. and 10 P.M., respectively, the use of Eastern Time would add over two hundred hours of sunlight a year.

The time zone campaign was waged in Cleveland over the next few years, until finally, in 1914, the Cleveland City Council approved a municipal ordinance to adopt Eastern Time. On the night of April 30, 1914, Clevelanders got one hour less sleep but enjoyed the unusual privilege of awaking in a new time zone. Local railroads, having had no say in the matter, continued using their own time zone plan, which still placed Cleveland in the Central time zone. Though the shift resulted in some inconvenience for travelers, who had to adjust their watches to Eastern Time upon disembarking at Cleveland's railroad terminal, the general feeling in the city was positive.

Meanwhile in Detroit, where the push to adopt Eastern Time had originated, opponents continued to fight against a time zone shift. Even after the shift in Cleveland, the *Detroit Free Press* persisted in its editorial opinion that a local shift would be "a profound mistake . . . causing much confusion." Nevertheless, George Renaud and the More-Daylighters continued to press for Eastern Time.

Among other means of persuasion the group undertook, it tried to get the backing of prominent local citizens, including those related to two of the passions of Detroiters, automobiles and baseball. Detroit was the "Motor City," the

CLEVELAND CITY ORDINANCE

Be it ordained by the Council of the City of Cleveland, State of Ohio:

Section 1. That the standard of time throughout the City of Cleveland shall be the seventy-fifth meridian of longitude west of Greenwich, known as "Eastern Standard Time." Municipal offices and legal or official proceedings of the City of Cleveland shall be regulated thereby; and when by ordinance, resolution or action of any municipal officer or body an act must be performed at or within a prescribed time, it shall be performed according to such standard of time.

Section 2. When a clock or other timepiece is on or upon a public building maintained at the expense of the City of Cleveland, the board, commission, officer or other person having control and charge of such building, shall have such clock or other timepiece set and run according to the standard of time as provided in Section 1 hereof.

Section 3. This ordinance shall take effect from and after the 30th day of April 1914.

Cleveland gets year-round daylight saving time by changing time zones.

automobile manufacturing capital of the country. The number of automobiles in use was surging, as the popularity of Henry Ford's Model T made them more available to the average person, who was coming to see the ability to take a drive in a car as part of the American way of life. Auto enthusiasts appreciated that an extra hour of daylight allowed more time for sunlit evening drives in the country, and Detroit car manufacturers valued the savings on lighting for their plants.

When Detroiters weren't talking about cars they were likely talking about their Detroit Tigers baseball team. The Tigers had a promising roster, led by the all-time baseball great Ty Cobb, who by 1914 had won the league batting title eight years in a row. Because baseball became more difficult to play as the sun went down, and games were often discontinued on account of darkness, an extra hour of daylight would be most welcome. As part of their campaign the More-Daylighters gathered long lists of endorsements by local luminaries, making sure to include prominent automobile executives and baseball stars. For example, the Packard Motor Car Company's general manager, Alvan Macauley, declared, "I am glad to express my approval of the objects and aims of the More-Daylight Club"; Hugh Chalmers, the president of Chalmers Motor Company (forerunner of Chrysler Corporation), was "heartily in favor of the movement to have all the daylight we can"; and Ty Cobb enthused, "Lovers of baseball should welcome a change to Eastern Standard Time."

Finally, a year after Cleveland acted, the Detroit Common Council, noting the success of Cleveland's time zone shift, finally succumbed to the efforts of Dr. Renaud and the More-Daylight Club and passed an ordinance adopting Eastern Standard Time, effective May 15, 1915.

THE KAISER'S TRICK HOUR

The time zone shifts in Cleveland and Detroit proved popular. In September 1916, the future of Eastern Time in Detroit was put to a citywide referendum and was retained by a vote of more than two to one. Groups sprang up in several other cities espousing some form of time advancement. There was also a fledgling movement to extend the Eastern time zone so it would encompass the whole of the United States east of the Mississippi River, and do away with the Central time zone completely. The Chicago Association of Commerce and Cincinnati's E. H. Murdock voiced support for such a plan. But except for a few small communities in the Detroit and

Cleveland areas, no other cities or towns chose to change their time zones.

Instead, interest in summer daylight saving time began to grow. As in Britain, initial proposals for DST often met with ridicule. People who advocated DST were called "nuts," "faddists," and "impractical dreamers." But by 1916 summer daylight saving time was being advocated for New York City by its Merchants' Association, for Chicago by its Association of Commerce, and by similar groups in Baltimore, Boston, Philadelphia, Kansas City, Providence, and elsewhere.

When Germany instituted DST in 1916, many DST advocates in the U.S. were heartened by its adoption in a major industrial country. But other Americans viewed the German plan through the prism of the ongoing European war and their feelings toward Germany. The *New York Times* characterized daylight saving time as "the Kaiser's Trick Hour," and the economist Joseph French Johnson, while personally favoring the concept of advanced time, couldn't help but note the difference in national characters: "The disposition of the American people is such that they would scarcely obey a mandate of Congress or Legislatures to change their watches if they didn't want to. In Germany, of course, it is a simple proposition. The Kaiser orders it, and it is done."

As the war continued and other European countries followed Germany's lead, opinion in America remained mixed. Many treated the subject lightly. A Bostonian suggested that the truly popular thing to do would be to have the clocks run an hour slow in the early morning and an hour, or perhaps two, fast in the afternoon. The *Saturday Evening Post* dealt with DST a little more jocularly than even that magazine's founder, Benjamin Franklin, might have liked, coming out in favor of the daylight saving plan because it was "one of those harmless pieces of buncombe which please many people while really hurting nobody. . . . There's no good reason to stop with the clock either. Why not 'save summer' by having June begin at the end of February?"

With public interest slowly growing, Congressman William Borland of Missouri and Senator Jacob Gallinger of New Hampshire sponsored a national daylight saving time bill in 1916. Supporting DST at a Senate hearing, John O'Laughlin, Washington correspondant of the *Chicago Herald*, appealed to the senators, all living in the Washington, D.C., area, by noting that the previous day the "baseball club which represents the National Capital and the one which represents Chicago battled 14 innings to a tie. Darkness prevented [your] young men from adding a much needed ball game to the winning column. With an extra hour of sunlight, Washington might well have been leading the league again." Nonetheless, the DST measure failed to pass.

THE CLOTHIER, THE CONDUIT MAKER, AND THE MERCHANT

It took a New York City official, a Pittsburgh manufacturer, and a Boston retailer to organize American daylight saving time supporters into a national effort.

In New York, Marcus M. Marks was a clothing manufacturer, the longtime president of the National Association of Clothiers, and a philanthropist. He became an enthusiastic supporter of DST while serving a four-year term, beginning in 1914, as Manhattan's borough president. Feeling there was a need for a more organized approach to establishing DST, on May 29, 1916, Marks gathered together fifteen representatives of labor groups, merchant associations, and local government to begin a nationwide movement for summer DST. This group became the National Daylight Saving Association (NDSA), and it grew to become the most important single-issue DST advocacy group in the country.

Along with Marks, one of the most important advocates of daylight saving time in America was the Pittsburgh industrialist Robert Garland. Garland was founder and president of Garland Manufacturing Company, purveyors of electric conduits, and was very active in community affairs, serving on the Pittsburgh City Council, as president of the Pittsburgh Chamber of Commerce, and later as chairman of the War Resources Committee for Western Pennsylvania and West Virginia. Garland had heard of William Willett's plan, and after examining data on Britain's savings of fuel and kilowatt-hours he became strongly convinced of the benefits of DST. At a Pittsburgh Chamber of Commerce meeting in the summer of 1916, Garland made a proposal for national daylight saving time. Though the steel men and businessmen were at first skeptical, he quickly won them over and the Pittsburgh chamber became the first U.S. business organization to endorse national DST.

Meanwhile in Boston, department store magnate A. Lincoln Filene became a third leading proponent of DST in America. Filene, together with his brother Edward, had built William Filene's Sons Company into a major Boston department store with a number of branches. Known as a forward-thinking businessman and manager, he originated or implemented a number of significant innovations, including cycle billing, the first company charge card, and the fostering of a strong positive relationship between management and workers. Filene became interested in the potential benefits of daylight saving time and convinced the Boston Chamber of Commerce to set up a Special Committee on Daylight Saving Time, on which he served as chairman. His committee studied the subject in detail and issued a report, "An Hour of Light for an Hour of Night," which recommended DST nationwide, pointing out its advantages and answering the objections against it. Among other arguments, the report pointed out that industrial accidents increased and efficiency decreased when artificial lighting was used. "Every unnecessary hour under artificial light means a direct loss of production . . . [and] the substitution of a cool morning hour for a hot afternoon hour raised worker efficiency." Although Filene's committee recommended a year-round DST plan, it also suggested that if a shorter period were to be used, it should be from April 1 to November 30.

Through their combined efforts, in 1916 Garland and Filene, with support from Marks's New York group, convinced the United States Chamber of Commerce to set up an eight-member National Committee on Daylight Saving, with Garland as chairman, and Filene and six other leading businessmen from around the country as members. The committee met in December 1916 at the William Penn Hotel in Pittsburgh and produced a lengthy report strongly recommending nationwide summer DST. The report cited many of the benefits traditionally attributed to DST, including production efficiency and improved working conditions. It noted, too, reversing an argument used against DST in Britain, that with most of Europe now observing daylight saving time, an hour of overlap in business operations could be regained if America also adopted DST.

Labor organizations representing millions of American workers also got on board. In January 1917, the American Federation of Labor urged daylight saving time for "the conservation of time and opportunity for greater leisure and open-air exercise for the masses of the people." And the considerable weight of the U.S. Chamber of Commerce's half million businessmen was put squarely behind DST when the chamber's

national convention in Washington, D.C., endorsed the recommendation of its National Committee on Daylight Saving
on February 1, 1917.

The movement was catching on. In early 1917, over one thousand enthusiastic DST advocates gathered at the Hotel Astor in
New York City for a two-day National Daylight Saving Convention organized by Marks's National Daylight Saving Association.
Representatives of government, industry, labor, and interest
groups from across the United States attended, all wearing
lapel buttons showing Uncle Sam turning the clock forward
one hour. Speakers included John Tener, president of the
National League of Baseball Clubs, New York's Mayor John
Mitchel, and Representative Borland of Missouri. A
spokesperson for the blind made a plea for an additional hour
of daylight for those whose eyesight was poor. A member of the
British House of Commons, J. H. Whitehouse, spoke about the

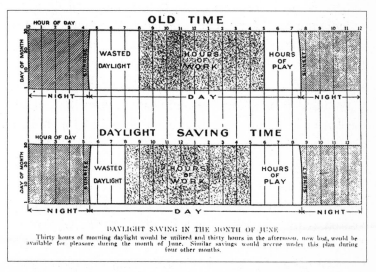

Chart used by DST advocates: Thirty More Hours of Light in the Month of June.

positive effects of summer time in Britain, and President Woodrow Wilson sent a letter of support in which he stated that he would be "glad to back up any movement which has the objects of the daylight saving movement."

But not everyone shared the conventioneers' pro-DST feeling. Content with the time zone system that they had already created, the railroads expressed concern about possible dangers arising from repeated clock changes, and in February 1917 the American Railway Association approved a resolution stating that "any legislation excepting that which provides for a permanent change and which recognizes the present standard time zones shall be opposed on behalf of the railways."

AMERICA AT WAR

With the United States inching closer to joining the war in Europe, Robert Garland, the DST leader from Pittsburgh, called for daylight saving time as "one of the most effective steps that can be taken towards 'Preparedness,' which is the order of the day." But still no national legislation was enacted. Then, on April 6, 1917, the United States, provoked by over two months of unrestricted submarine attacks against American ships, declared war on Germany. The United States had maintained its neutrality for almost three years, but enmity escalated when British intelligence intercepted a telegram from the German foreign secretary, Arthur Zimmermann, to Mexico in which Zimmermann proposed an alliance with Mexico against the U.S. and promised Mexico some American territory in exchange for backing the German cause. Finally President Wilson, with full support from his cabinet, decided that the United States could not continue to tolerate the unprovoked German submarine attacks and asked Congress for a formal declaration of war.

Just as the war had driven the acceptance of daylight saving time throughout Europe, it similarly energized the movement in America. Less than two weeks after the United States entered the war, a bill drafted by the National Daylight Saving Association "to save daylight and to provide standard time" was introduced in Congress by Senator William Calder, a Brooklyn Republican, and by Representative William Borland, a Kansas City Democrat. The Calder-Borland bill called for five months of daylight saving time, from May through September, and also provided for finally making the railroads' standard time zones official.

To apply pressure, the National Daylight Saving Association organized a massive postcard campaign to members of Congress. Leaders of the DST campaign, including Marks, Garland, Filene, Renaud, and Professor Harold Jacoby of Columbia University, testified before Congress in favor of the bill. Marks focused on practical benefits to the war effort, such as how an extra hour of home gardening would increase food production. A Harvard University astronomy professor, Robert Willson, presented a study of cities situated near the railroads' time zone boundaries, which showed that a great majority of those cities had chosen to be part of the more easterly of the two time zones, gaining an hour, and thus "a great many people have been saving a good deal of daylight for years." Others who testified concentrated on fuel conservation and health improvement, and produced a long list of supporters, including 150 daily newspapers, the National League of Professional Baseball Clubs, the National Lawn Tennis Association, the American Federation of Labor, and President Wilson.

Providing the businessman's point of view, Sidney Colgate, president of Colgate & Company of Jersey City, New Jersey, testified about the effect his soap, toothpaste, and perfume

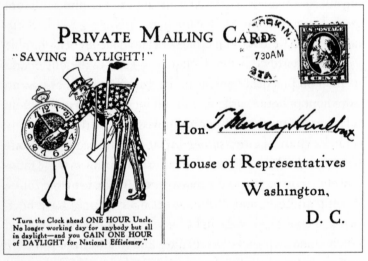

June 4th 1917

Hon. *T. Murray Hulbert* M. C.
Washington, D. C.

Dear Sir:-

If I have more Daylight I can work longer for my country. We need every hour of light. As a voter in your district I favor the immediate passage of the Daylight Saving bill by Congress.

Name *G. F. Haskell*

Address *320 St. Nicholas Ave*

Town or City *New York City*

State *N.Y.*

PRIVATE MAILING CARD

"SAVING DAYLIGHT!"

730AM

"Turn the Clock ahead ONE HOUR Uncle. No longer working day for anybody but all in daylight—and you GAIN ONE HOUR of DAYLIGHT for National Efficiency."

Hon. *T. Murray Hulbert*

House of Representatives

Washington,

D. C.

Postcards to Congress urging DST.

manufacturing company had already experienced with a change in the hours of daylight. In 1915, the company had voluntarily shifted its summer business hours one hour earlier. "We tried it beginning with the middle of July," he recounted, "and we thought we would try it until the first of September.

Toward the end of August we took a vote as to whether we should continue it to the first of October, and we got a 94 percent vote to continue." That fall he recalled pointing out to one department head that he had not yet had a vacation, to which the man replied, "I do not feel as if I [need] any vacation. Every day during the summer was . . . a short vacation, and I have never been in better health." As a result of the program, Colgate found that company employees worked with much more "snap and energy."

D. C. Stewart of the American Railway Association spoke out against the bill, citing the risk of accidents on single-track railroads and noting that there were 1,698,818 clocks and watches on all the railroads that would need to be changed. A shift to DST presented a dangerous situation—any confusion could result in catastrophe. "If there should be one man . . . on any railroad who misinterprets his instructions," he argued, "or if he makes a mistake in changing his watch—if he wants to change it an hour and he does not change it quite an hour, or if he wants to change it forward and inadvertently changes it back, there is nothing that stands between that man and a wreck."

To minimize potential problems for the railroads, Senator Frank Kellogg of Minnesota asked Stewart at what point in the day the fewest number of trains were running. Stewart told him two o'clock in the morning, thus confirming 2 A.M. as the designated transition time in this bill—and in all American DST laws since.

The Senate passed the Calder-Borland bill on June 27, noting its "special value in the present national emergency," and the focus of DST efforts shifted to the House of Representatives. Harry A. Garfield, the government fuel administrator, issued a plea for passage, describing the importance of the measure in the conservation of fuel, and Food Administrator Herbert Hoover strongly endorsed "this excellent bill" as a way to increase food production. The Shipping Board backed the bill and so did P. S. Risdale, president of the National War Garden Commission. Risdale projected that DST would add an estimated 910 million hours of home gardening each year, producing vegetables that could be substituted for enough meat to feed an army of one million soldiers for six months.

The United Cigar Stores Company sponsored a campaign promoting the wartime benefits of DST, producing a series of posters that appeared in the windows of more than two thousand cigar stores on street corners across the country. Most major newspapers and magazines had become backers of DST, and the leading popular science journal, *Scientific American,* previously hesitant, was now unequivocal in its support of the measure:

A United Cigar Stores daylight saving campaign poster.

"Under the stress of the war," it editorialized in January 1918, "European countries have put this scheme into effect with the result that there has been an enormous saving of fuel for lighting purposes. The value of [the] measure is not to be debated."

As support for the bill grew, farmers and railroad representatives maintained their strong opposition, and there were other unenthusiastic views. Congressman Otis Wingo of Arkansas mused, "I do not know that I have any particular objection to this bill; I just decline to take it seriously. . . . A majority of the men who advocate this character of legislation have not seen the sun rise for twenty years. . . . This bill is for the relief of the slackers of the nation who are too lazy to get up early." Echoing a 1909 taunt that appeared in the journal *Nature*, he suggested that Congress go on to provide special thermometers with the freezing mark labeled 45° F instead of 32° F. This would save coal, since people would leave their home thermostats at 70° and be satisfied with thirteen degrees less heat. Wingo concluded on a more serious note: "We should not be wasting our time on such bills, but should go on to the war-finance bill. . . . While our boys are fighting in the trenches, we are here like a lot of school boys 'tinkering' with the clocks."

Several representatives noted that more coal was consumed in March and October than in any two of the months falling under the scheme, and the House amended the Senate bill to extend the DST period to seven months, from the last Sunday in March to the last Sunday in October. The amended bill was passed in the House on March 15 by a vote of 253 to 40. The Senate approved the amended bill the next day, and President Woodrow Wilson signed it into law on March 19, 1918. It was called the Standard Time Act of 1918, and the part of the act providing for DST was called the Daylight Saving Time Act.

CELEBRATIONS AT 2 A.M.

The Standard Time Act established seven months of national daylight saving time each year, April through October. It also legally sanctioned the railroads' four time zones—Eastern, Central, Mountain, and Pacific—and added a new fifth zone, Alaska Time, for all of Alaska. The Interstate Commerce Commission (ICC), the only federal transportation regulatory agency at the time, was authorized to adjust time zone boundaries and to govern time regulation but was given little power to enforce time standards. Legally, the time zones applied only to federal government operations and to the movement of common carriers engaged in interstate or foreign commerce. But although the act did not mandate a legal time for any other purpose, it was clear that federal legal time—based on the time zones plus DST when applicable—would be the basis of time for most other pursuits.

65th Congress, Public Law 65–106

An Act to Save Daylight and to Provide Standard Time for the United States

Be it enacted by the Senate and House of Representatives of the United States of America in Congress assembled,

Section 3. That at two o'clock antemeridian [A.M.] of the last Sunday in March of each year the standard time of each zone shall be advanced one hour, and at two o'clock antemeridian of the last Sunday in October in each year the standard time of each zone shall, by the retarding of one hour, be returned to the mean astronomical time of the degree of longitude governing said zone, so that between the last Sunday in March at two o'clock antemeridian and the last Sunday in October at two o'clock antemeridian in each year the standard time of each zone shall be one hour in advance of the mean astronomical time of the degree of longitude governing each zone, respectively.

Approved March 19, 1918

The Standard Time Act of 1918, Daylight Saving section.

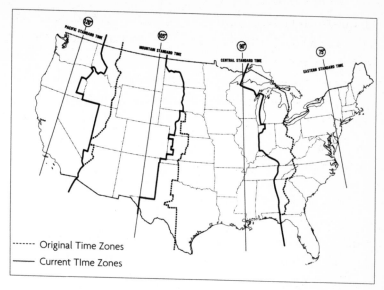

Original official U.S. time zones.

America's first daylight saving time period began at 2 A.M. on Sunday, March 31, 1918, less than two weeks after passage of the Standard Time Act.

The National Daylight Saving Association urged factories, shops, offices, and plants that would not be open Sunday to set their clocks ahead an hour before closing on Saturday. Workers on the Saturday midnight shift were told to expect to finish at 9 A.M. instead of 8 A.M. Pastors worried that many parishioners would be an hour late for Easter services, and officials at the NDSA suggested that "churches ring their bells more lustily than usual" that Sunday morning.

Railroads announced that trains between terminals at 2 A.M. would arrive one hour late but start out again on the new time. Frank Hedley, general manager of New York City's mass transit line, issued a notice stating that at precisely 2 A.M. an "alarm will be sent from the trainmaster's office on all the subway and elevated divisions, which will be an indication to all dispatchers, station agents, and others to immediately advance their watches and clocks exactly one hour."

On the day before the change to daylight saving time, the *Chicago Tribune* reminded its readers with a couplet by "Little Mary":

> TONIGHT'S THE NIGHT
> Just before you go to bed,
> Push the clock an hour ahead.

And "B.T." contributed another *Tribune* couplet, touching upon what some believed to be the real purpose of the time change:

> THESE BUSY DAYS
> Forward, turn forward, O Time in your flight!
> So we can squeeze in nine holes before night.

B.T. added waggishly:

> Of course you know what time the 2:15 train goes out Sunday
>   morning.
> Neither do we.

On the night of the time change, families all over the country turned their clocks ahead one hour. Many sat up until two o'clock Sunday morning to do so at the official time. "Your honest-faced old clock," the *Chicago Tribune* reflected, "will look at you straightforwardly and blandly inform you that it is 6, 7, 8 or 9 o'clock and you will take him at his word because you know Uncle Sam is backing him, but way down deep in your heart, for a day or two at least, you will not be able to forget that the old fellow is living a lie—a lie that will benefit you and your country."

In many cities across America, veterans of the DST campaign held celebrations at 2 A.M. Marcus Marks and members of the NDSA met at the Aldine Club in New York to celebrate the advancement of the clock on the Metropolitan Tower on Madison Square. A large outdoor celebration took place simultaneously in nearby Madison Square park. Several thousand spectators watched a parade by the Boy Scouts headed by the New York Police Department Band and listened to patriotic addresses by War Saving and Liberty Loan speakers. Surrounded by enthusiastic DST supporters, Marcus Marks pushed a button that illuminated the big Metropolitan clock, which had been dark for an hour, and moved its minute hand one complete revolution, as the clock advanced from 2 A.M. to 3 A.M. A delegation of men from labor unions then presented Marks with a clock in recognition of his work for the new law. In nearby Brooklyn, Senator William Calder was the principal speaker at another patriotic rally at Borough Hall. At 2 A.M. Calder moved the hands on a small clock forward as the building's large clock was advanced by an hour.

Ironically, Detroit, one of the first leaders in the campaign for more daylight, served as a kind of "conscientious objector" during the festivities. Detroit's switch to the Eastern time zone in 1915 had given it year-round daylight saving time, and the home of the More-Daylight Club decided it didn't need any more daylight. So the day before national DST went into effect, the Detroit City Council approved an ordinance making Central Time the official time in Detroit during the national DST period, in effect keeping its time exactly as it was. Cleveland, possibly because it lies a little more east than Detroit, did not enact such an ordinance and moved its clocks ahead on March 31 along with the rest of the country.

MEANWHILE, IN CANADA

Well before World War I and even before Renaud and Hayes formed the More-Daylight Club in Detroit, a Canadian by the name of John Hewitson saw the benefit of a time zone shift in climes farther north. Sometime around 1904, Hewitson, a young teacher in the village of Rossport, Ontario, moved one hundred miles west to the small shipping community of Port Arthur, on the northern shore of Lake Superior in the western reaches of Ontario.

Hewitson was a well-liked young man with a disposition that radiated optimism—it was said that "a five-minute talk with him was a tonic." In Port Arthur, Hewitson became a successful businessman and later founded his own construction company. He was an enthusiastic advocate of outdoor sports in the Port Arthur area. He realized, with reasoning similar to the More-Daylighters, that by switching the time zone of Port Arthur from Central Time to Eastern Time, residents would have longer evenings of light. He was particularly interested in the additional time for athletics that an extra hour of daylight would bring; as a former teacher he knew that students would greatly benefit from an extra hour of evening exercise.

Hewitson proposed a permanent time zone switch for the town of Port Arthur. He gained the support of the local board of trade and managed to enact his plan in a much shorter time than had either Willett or Renaud; of course, Port Arthur was a much smaller jurisdiction. In 1908, the Port Arthur Council passed a bylaw switching the town to Eastern Time for a two-month summer trial. Although they thought of it as a temporary shift of time zone, Port Arthurians may well have been the world's first beneficiaries of summer daylight saving time.

The summer trial proved so successful that the next year Hewitson and many others called for Port Arthur to institute Eastern Time year-round. Over a smattering of objections, a vote was taken and Port Arthur made the switch permanent on April 30, 1909. Having observed the success of Port Arthur, the citizens of nearby Fort William followed in their neighbors' footsteps and made the change in 1910. Now the two principal communities of Ontario's Lakehead Region were permanently on Eastern Time. That, however, was as far as Hewitson's proposal spread.

Canada, a dominion of the British Empire, followed the daylight saving controversy in Britain as Parliament debated Willett's proposals. In 1909, a national DST bill was introduced in the Canadian Parliament, but it met a similar fate as Willett's many attempts. With no national law, it would be up to cities themselves to experiment with daylight saving time.

One such city was Halifax, Nova Scotia. On April 17, 1916, the city council of Halifax unanimously approved a measure to initiate a "daylight saving scheme" for that city from May 1 to September 30. The board of trade president, G. W. Henley, urged advocates of temperance to endorse the scheme as a way to minimize wrongdoing: "for there would be more daylight in which to see what was going on." But a worker raised the objection that Halifax tramway employees and others required to rise at 4:30 A.M. had been "waiting all through the long dreary months of winter for the time to come when they may see a glimmer of daylight before going to work." In contrast, he noted that among the backers of DST were "men whose business hours permit them to sleep until eight in the morning."

Our mission is to serve well.

Each of Our Customers Will Please Take Notice

that we have fallen in line with THE DAY-LIGHT SAVING SCHEME, and in accordance therewith our Warehouse from May 1st to September 30th will

Open at 7 O'Clock a. m., and Close at 5 p. m.
OR, EASTERN STANDARD TIME
(now in operation)
Open at 6 O'Clock a. m., and Close at 4 p. m.

We would kindly ask our customers in the Provincial Towns where THE DAYLIGHT SAVING SCHEME has not yet been adopted to have their telegrams ordering goods for shipment by early morning Express reach us before 4 O'CLOCK EASTERN STANDARD TIME.

WM. STAIRS, SON & MORROW, LIMITED
ESTABLISHED 1810
EASTERN CANADA'S SUPPLY HOUSE,
HALIFAX

Companies tried to notify their customers of local DST.

The next meeting of the Halifax city council was packed to overflowing, and the assembly was interrupted numerous times by the jeers, applause, shouts, and laughter of proponents and opponents. The city council vigorously reaffirmed its previous decision, and the daylight saving scheme went into effect in Halifax on May 1, 1916. Although a majority of people seemed pleased with the decision, a reader who signed herself "Working

Woman" wrote to the *Halifax Herald*, "It is very easy for the class of people who do not know what hard work or early rising really means to take up this fad." A letter from a "Workingman" concurred that rather than a daylight saving scheme, "it would more appropriate to call it the 'moonshine scheme.'" Under the new system, the worker would have to be up extremely early. "And for what? To enjoy an hour longer in the evening! Believe me, as a rule when a man has done a fair day's manual labor, he is glad to smoke his pipe, read his daily paper, and go to rest."

Proponents and opponents debated the merits of the new system and both launched petitions signed by thousands of citizens. When the measure came up for renewal the following year, opponents pointed out that Halifax's purely local DST had isolated it from the rest of Nova Scotia, and Halifax decided not to reinstitute it.

In 1918, after the passage of the Daylight Saving Act in the United States, Canada remained on standard time while it debated a DST bill. When the Canadian House of Commons defeated the bill, the Canadian Railway Board brought the issue to a head by declaring that if no bill was passed, the railroads would, on their own, follow the American DST plan. Once DST had started in America, Canadian trains going to the United States had to stop and wait an hour at the border so as to keep to published schedules. As a result of this declaration, the Canadian Parliament reconsidered and passed the Daylight Saving Act of 1918, and Canadian DST was initiated on April 14, 1918. Canada's time once again corresponded to that of the United States, which had started DST two weeks earlier.

GET UP EARLY TO BEAT THE KAISER
On the first day of national DST in the United States, the *New York Times* reported, "There were plenty of baseball games,

and, so far as could be learned, no games were called 'on account of darkness.' . . . Those who looked toward the sky at nightfall expecting to see a miracle were disappointed. The sun did not stand still at 6 o'clock, but there was still sufficient light to read by and to play by until 7 o'clock." Many arrived late for morning meetings and services and a few insisted on sticking to the "old time," but on the whole the time change was generally observed. TIME CHANGE BIG SUCCESS was the headline of the *Los Angeles Times*'s article. "The inauguration of the daylight-saving law yesterday was not fraught with the disasters that many people thought would attend the changing of time all over the country. . . . Things moved along uninterruptedly, no great amount of inconvenience being caused, and no mishaps resulting."

At Camp Dix, in New Jersey, there were some complaints when "Taps" sounded one hour early, and army cooks reported that their companies ate 30 percent less when summoned to lunch at the equivalent of eleven o'clock. Around the country, merchants reported that on the first business day of DST there was no need for artificial light at the end of the day, and New York Stock Exchange members were pleased to have an hour of overlap with the London Stock Exchange. Motorists, most of whom preferred driving in daylight, had an extra hour of Sunday touring time.

The task of resetting the clocks throughout the nation caused some difficulties. At the Elgin National Watch Works, in Elgin, Illinois, changing the time on the huge main clock, twenty feet in diameter and with a five-hundred-pound pendulum, required three men working for an hour. Dr. Henry Cox, chief of the U.S. Weather Bureau, pointed out that weather records would be kept according to "summer time" as well as "real time," giving his office double duty. But the overall mood was positive toward

an hour is added

on the clock today

years are added to
any life

by making a Maxwell
car meet the demands

on time and energy

a Maxwell can be bought for
$275 down and the balance
monthly up to $825

President

Harry Newman - Stratton Company
Michigan Avenue at Twenty-fifth Street
ALSO

Dewitt Smith Bldg.,   435-439 Main St.,
Springfield, Ill   Peoria, Ill

American advertisers were quick to take advantage of DST.

daylight saving time, at least as a wartime measure. The *Literary Digest* captured this feeling: "We are all going to get up early to beat the Kaiser. This, in brief, is the meaning of the Daylight Saving Law." In Massachusetts, the *Springfield Republican* called it "frankly a war-measure, in line with the early closing of places of amusement."

As daylight saving time continued in the United States

through the summer of 1918, it met with approval from the majority of American newspapers and magazines, and much of the public as well. Senator Calder exuberantly declared, "Without question this bill has been more helpful in the great war-work in which this nation has been engaged than any other one thing," and the *Literary Digest* reported the national consensus that "daylight saving is now a permanent feature of our national life." Some advocates in Congress even tried to extend DST to the entire year, pointing out that in winter it would reduce the peak load of power plants and save more fuel. But the year-round extension did not pass, and the first period of DST in the United States came to a close on October 27, 1918.

DAYLIGHT SAVING TIME SNAPSHOT: WORLD WAR I	
Africa	Portugal
Algeria	Russia
Tangier	San Marino
	Spain
Europe	Sweden
Austria-Hungary	Switzerland
Belgium	Turkey
Denmark	United Kingdom
Estonia	
France	**North America**
Germany	British Honduras
Iceland	Canada
Italy	Newfoundland
Latvia	United States
Luxembourg	
Malta	**Oceania**
Monaco	Australia
Netherlands	
Norway	**South America**
Poland	Chile

DST during the years of World War I.

In Britain, France, Germany, and most other countries that had adopted it for wartime purposes, DST continued through the turbulent summer of 1918. The Germans launched a major offensive in March 1918, but British and Empire forces eventually pushed them back. Then the Allies, with the help of the recently arrived American Expeditionary Force, went on the offensive and claimed a series of victories, ultimately leading to requests for armistice by each of the Central Powers. At the eleventh hour of the eleventh day of the eleventh month of 1918, the war guns fell silent. In a railroad car in Compiègne, France, Germany signed an armistice with the Allies and the Great War was over.

In the aftermath of the war, after years of fighting and the loss of some nine million lives on the battlefield and about the same number of noncombatant lives, the countries that had been involved wanted nothing so much as to get "back to normal." And one of the issues facing them was the question of what should be done about that peculiar wartime measure, daylight saving time.

Chapter Five

An Hour of Peace

Rise with the lark, and with the lark to bed.
—James Hurdis, *The Village Curate*

*B*y the time the Great War ended in November 1918, more than sixty-five million soldiers had been mobilized, and fifteen million people had lost their lives. Now the world's thoughts turned to peace and normalcy. Millions of soldiers returned home to civilian life, and factories that had been retooled to manufacture weapons resumed the production of consumer goods. Wartime restrictions were relaxed. Even though many had supported daylight saving time during the war, each country that had adopted it now had to decide what to do next.

THE FARMER VERSUS THE CITY MAN

In America, daylight saving time had been observed from April through October of 1918. When the war ended, President Wilson and many Americans thought DST should continue in future years, but many others saw it as a wartime measure like "Meatless Mondays" and "Wheatless Wednesdays," and it had always been unpopular with certain groups, especially farmers. Congress began to feel increasing pressure to repeal the daylight saving provision of the Standard Time Act. The National Daylight Saving Association urged Congress to keep the law, and Samuel Gompers and other leaders of the American Federation of Labor issued statements backing it, as did most politicians and newspapers from large cities, especially in the East. The opposition was led by farm organizations, such as the National Grange, and by politicians and newspapers representing rural areas. As the *New York World* characterized the controversy, the battle over DST had "boiled down to a contest between rural and city workers."

TRYING TO TAKE THE BULL BY THE HORNS.

Farmers wrestled with DST.

Why were farmers so strongly opposed to daylight saving time? Since farmers worked by the sun, it might be argued that the shifting of the clock did not directly affect them—they would still rise before daybreak and work during daylight, regardless of what hour the clock said. But farmers traditionally made use of the discrepancy between their sun-time hours and the usual clock-time hours of the rest of the community to allow for commerce and leisure activities, and the introduction of DST destroyed that balance. The change had a powerful impact when farmers had to interact with the clock-oriented world—railroads, hired hands, schools, banks, businesses, stores, government offices, community activities, movie theaters—and that impact was usually negative.

Senator Arthur Capper of Kansas became a visible spokesman for the farmers' position. Through the newspapers he owned, such as Topeka's *Capper's Weekly*, he catalogued some of the farmers' principal objections to daylight saving time. Under DST, farmers shipping milk or other perishables had to meet trains an hour earlier than they were used to. They had to leave their farms an hour earlier during the day to get to stores and banks in towns. They had to quit work an hour earlier in the evening if they wanted to attend community gatherings or other activities. Their seasonal hired hands insisted on working on "town time"—starting work when it was still hours before the dew had evaporated and hay and grain could be bundled and baled, and then quitting work hours before sundown. In effect, DST had added one hour of darkness to the farmer's day instead of one hour of daylight.

While many urban DST supporters scoffed at the farmers' objections, others found them reasonable and sought means of resolving them. A large portion of the grievances could "be removed by comparatively simple adjustments," which "should be hurried," wrote the *New York Evening Post*. "Can not the Railway Administration and the Labor Department, respectively, do something to bring the hours of milk trains and of farm labor nearer the farmers' demands? If we can make the daylight saving system work more easily in rural districts, the clamor for repeal will die away."

But no such measures of conciliation were implemented, and in early 1919 farm representatives in Congress introduced over twenty bills to repeal the daylight saving law. John Esch of Wisconsin, chairman of the House Commerce Committee, introduced his own repeal bill, which became the focus of debate. A letter to Congress signed by a group of nearly one hundred Missouri "farmer voters and taxpayers" declared that the daylight saving plan "from the standpoint of the vast agricultural interest of the Nation . . . has proved a 'daylight-wasting' plan." Thomas Atkeson of the National Grange, which represented 750,000 dues-paying agriculturalists, strongly expounded the farmers' case against DST: "We loaf around an hour in the morning waiting for the dew to dry off. . . . Our men quit at 5 o'clock in the afternoon, and they have done maybe six hours' work when they ought to have done eight. . . . If you want to cut off 25 per cent of the productiveness of the American farmer, just keep this law on the statute books."

As the calls for repeal grew stronger, charges and counter-charges flew. DST supporters claimed it was the electric companies, concerned about withering profits due to decreased use of artificial light, who were really behind the repeal movement. On the other side, Representative Edward King of Illinois called for the repeal of "the insane piece of legislation known as the Daylight Saving Law," and asserted, "The charge that electric-light companies are in back of the movement for repeal of the law is a simple, plain, unvarnished falsehood. The demand for repeal is universal."

Some considered DST a perversion of the natural order of things. "Repeal the law and have the clocks proclaim God's time and tell the truth!" exclaimed Mississippi Congressman

Ezekiel Candler Jr. "When we passed the law, we tried to 'put one over' on Mother Nature," Harry Hull of Iowa observed, "and when you try to improve the natural laws it usually ends in disaster."

And in the manner of those who favored conspiracies, Representative Edward King of Illinois cited the influences he believed were behind support for DST: the New York Stock Exchange, the manufacturers of gardening and canning equipment, and the "professoriate"—"the impractical, the Utopians, the theorists, political astrologers, medicine men, and the advance agents of the millennium."

On the pro-DST side, Pennsylvania Congressman William Vare used a song by Harry Lauder, a well-known British music hall singer, to point out the benefits of utilizing morning daylight.

> O it's nice to get up in the mornin'
> When the sun begins to shine,
> At 4 or 5 or 6 o'clock in the good old summer time!
> When the snow is snowin'
> And it's murky over head,
> O it's nice to get up in the mornin',
> But it's nicer to lie in bed.

In line with the rural-versus-urban split, New York's Representative Fiorello La Guardia called the Daylight Saving Act "a blessing to the people of this country," and New Jersey's Ernest Ackerman linked the continuation of DST to the end of the war: "Those who have gone to the front and are now coming home to us, . . . are they not deserving of the extra hour of recreation this bill [for repeal] will deny them?" Despite these impassioned words, when the debate concluded the House of Representatives had passed House Commerce

Committee Chairman Esch's repeal bill by a strong majority, 232 to 122.

Meanwhile, in the Senate, opponents tried to defeat DST using a legislative maneuver whereby a rider for a controversial measure is attached to a completely unrelated but popular bill, with the hope that the whole bill will be approved. They added a clause repealing the Daylight Saving Act to the annual appropriations bill for the Department of Agriculture, legislation that commonly sails through Congress. The tactic worked—both the Senate and the House of Representatives passed the bill with the DST repeal and sent the legislation to President Wilson for his signature.

But repeal would not be so easy. On July 12, 1919, Wilson refused to sign the agriculture bill, vetoing it for the sole reason that it contained the DST-repeal rider. "I believe that the repeal of the act would involve a serious economic loss," argued the president; the law "served the daily convenience of the many communities of the country in a way which gave all but universal satisfaction." With public interest high, anti-DST forces in Congress declared that they would attempt to enact the repeal by overriding President Wilson's veto, a move requiring the votes of two-thirds of the members of both houses of Congress.

Both sides rallied. Marcus Marks, still president of the National Daylight Saving Association, exhorted his followers, "Rain, by telegraph and letter, demands upon Senators and Representatives in Congress to uphold the President's veto. Millions demand its retention, but the only way to make it known that 90 percent of the people want daylight saving is by protest and petition to Congress." But farmers and other opponents of DST flooded Congress, newspapers, and magazines with petitions, telegrams, and letters as well. "God knows more about time than President Wilson does," snapped one irate letter writer from Kansas City. On July 14, when the House of Representatives met for a vote, 247 members voted for the override and 135 voted against—just shy of a two-thirds majority. A switch of only eight votes would have changed the outcome, but for now, the daylight saving law had withstood another onslaught.

Nonetheless, the opponents of daylight saving time persevered. Chairman Esch's repeal bill, having been passed previously by the House, was now approved by the Senate, and for the second time in a month legislation to repeal DST was sent to Wilson for approval.

In a rare move, the president used his veto power on the same measure a second time. This time, aware of the very close vote on the previous override attempt, Wilson struck a more conciliatory tone. He vetoed this second bill "with the utmost reluctance. I realize the very considerable and in some respects very serious inconveniences to which the daylight saving law subjects the farmers of the country. . . . But I have been obliged to balance one set of disadvantages against another and to venture a judgment as to which were the most serious for the country."

Once again, an effort to override the veto was mounted. This time the House was able to muster the two-thirds needed, voting 223 to 101 on August 19. The Senate overrode the president's veto the following day, by a vote of 57 to 19, and thus enacted the law repealing DST.

In this remarkable flurry of legislative activity, bills to repeal DST essentially were passed twice, vetoed twice, and overridden once, all within a little over a month. Throughout the campaign, loyalties were not divided along party lines; rather, as the *New York Tribune* observed, it was a case where "the farmer defeated the city man." Marcus Marks lamented, "The subtle, sinister propaganda waged against daylight saving was too powerful for a few to combat successfully." Under the new law, national advanced time would continue in 1919 until the end of the current DST period but would not be resumed in the spring of 1920. Expressing the regret of DST advocates everywhere, Senator William Calder, the DST law's original sponsor, predicted sadly, "This means the daylight saving is through forever. I cannot understand how people allowed it to be defeated. I doubt whether any Congress will ever pass it again."

66TH CONGRESS, PUBLIC LAW 66–40

AN ACT FOR THE REPEAL OF THE DAYLIGHT SAVING LAW

Be it enacted by the Senate and House of Representatives of the United States of America in Congress assembled, That:

Section 3 of the Act entitled "An Act to save daylight and to provide standard time for the United States," approved March 19, 1918, is hereby repealed, effective on the last Sunday of October, 1919, after the approval of this Act, when by the retarding of one hour the standard time of each zone shall be returned to and thereafter be the mean astronomical time of the degree of longitude governing each zone defined in section 1 of said Act approved March 19, 1918.

Enacted by a ⅔ vote of both houses on August 20, 1919.

The DST repeal act, overriding President Wilson's veto.

RAGTIME

After the national law was repealed, it was left to each individual city or state to determine whether or not it wanted to retain DST. Although DST had been repealed, the provisions of the Standard Time Act that had established national time zones continued in effect, and over the next several years, in response to local requests, Congress and the Interstate Commerce Commission repeatedly adjusted the time zone boundary lines.

These time zone changes were made for two primary reasons: to move an area into the same time zone as a major commercial center, especially New York or Chicago, and to provide more light in the evening. For both reasons, the time zone boundaries were almost always shifted to the west. Each westward move of a time zone boundary moved the involved regions one time zone to the east and provided them with the equivalent of year-round daylight saving time.

These shifts in time zone came to influence residents' views of daylight saving time itself. Residents of areas that had switched, say, from Central to Eastern Time felt they already had year-round daylight saving time. Therefore they fought against adding DST. Another hour would effectively advance their clocks two hours, yielding very late sunrises and a wide disparity between the clock and local sun time.

With no national DST legislation in place, throughout the country pro- and anti-DST forces clashed on the state and local levels. The push to resume DST was strongest in the Northeast. "Daylight saving by local option is not going to work without effort," declared the *New York Tribune*, "but the bother is little and the gain is great." The chief "bother," according to the *Literary Digest*, was the confusion caused by the use of two times for railroad schedules in the same city: local suburban trains on daylight saving time and long-distance trains on standard time.

GUMMING UP THE CLOCK.

Fight for control of the clock.

The sentiment expressed by the *Jersey Journal* of Jersey City was typical of the advocacy of local DST by newspapers in the East. "All the industrial centers of the State are . . . for its passage . . . and its beneficent influence will be felt in every walk of life." The *Journal* then considered the opposition's position: "Some slight readjustment of farming conditions may be required under daylight saving, but what of it? Is the farmhand to be petted and pampered to the disadvantage of the industrial worker, who in this State far outnumbers him? Is the tail to wag the dog in New Jersey?"

Daylight saving was especially popular in New York City, the home of Marcus Marks and the National Daylight Saving Association. In September 1919, only a month after repeal had been passed by Congress and well before it was to take effect, Alderman Robert Moran proposed an ordinance to continue New York City's daylight saving time, contending, "This repeal was effected through the mistaken effort of the farmer-employer, inspired and augmented by the avarice and selfishness of the lighting trusts throughout the country." Even when stricken with appendicitis and at the point of being rushed to Lincoln Hospital, Alderman Moran still took the time to stop and express support for his ordinance: "Only the selfish, fanatical, and those who refuse to get out of a lifelong rut can be opposed to daylight saving." The city's health commissioner, Dr. Royal Copeland, stated that he was supporting the measure on behalf of the 700,000 poor children in the city's tenements, for whom one more hour of sunlight would be a significant benefit, while the New York Times backed the proposal but called its position "a confession of hopelessness" as to the chance of ever reviving national DST.

On October 14, 1919, the New York City aldermen voted unanimously in favor of the DST ordinance, and Mayor John Hylan approved it on October 24. Similar ordinances were enacted in a few other locations, such as Pittsburgh, where Robert Garland, still a councilman, led the fight to get DST reinstated in December 1919.

Also in the Northeast, in 1920 the Massachusetts legislature approved a statewide DST measure by a large majority, and on April 9, Governor Calvin Coolidge signed a bill "to restore the benefits of 'Daylight Saving,' so-called," into law—making

Massachusetts the first state to adopt DST after repeal of the national law.

Few complaints about DST came from Boston-area residents, but in the rural parts of Massachusetts, many passionate voices were raised in opposition, blaming the law on "a few golf players in Boston." The opposition was led by Herbert Myrick, a publisher of several farm journals, who formed the New England Conference Against Daylight Saving Time and tried, unsuccessfully, to stop the law through court injunction. When the Boston & Maine Railroad announced it was changing its schedules to conform to Massachusetts time, fierce protests rang out throughout northern New England. According to New Hampshire's *Manchester Union*, most people in the region resented Massachusetts using the railroads "to force its obnoxious law upon the people of New Hampshire, Maine, and Vermont."

When the B&M characterized its decision as "the greatest good for the greatest number," northern New England dairy farmers, who used the trains to deliver their milk to the Boston market, rebelled against having to get up yet another hour earlier. Under the new measure, many farmers would have to start milking their cows at 1:30 A.M. Furthermore, the six thousand rural New Hampshire schoolchildren who used the milk trains to get to school would now have to leave for school an hour earlier and then wait around for classes to start. Governor John Bartlett of New Hampshire appealed to Governor Coolidge "to prevent the imposition of a great injustice on New Hampshire," but Coolidge stood his ground and daylight saving time came into effect in Massachusetts on April 25, 1920.

New York was the only other state with DST in 1920, having passed such a law during World War I. In subsequent years the use of DST proliferated, primarily in the North and East in cities such as Providence, Bangor, Hartford, and Newark, and around the country in scattered locations such as Cincinnati, Columbus, Kalamazoo, Pittsburgh, Chicago, Denver, and several others. Anti-DST forces were able to repeal New York's statewide DST law in 1921, although New York City's DST ordinance remained in effect.

From Willett's first proposal, one of the arguments most frequently used against daylight saving time was that it was not necessary to advance the clocks—all that was necessary was to shift the hours people arose and worked. Some companies had tried opening early on their own, but the concept had never been attempted on a large scale and backed by government authority. But in 1922 the idea received the backing of a very important authority—the president of the United States.

With national daylight saving time repealed, the federal government still retained control over DST in one location, the District of Columbia, where it had a mandate to run the affairs of the capital city. In lieu of a law passed by Congress, Washingtonians would have DST only if the president issued an executive order to move the clock ahead. And President Warren G. Harding refused to do so—he saw no reason for the "deception" of advancing the clock hands. Better, he asserted, would be a plan to save daylight by shifting the hours people woke up and went to work.

So President Harding sent out a request order to government departments to start work at 8 rather than 9 A.M., beginning May 15, 1922. When that day arrived, most federal employees reported to work an hour earlier, local transportation companies adjusted their schedules, and most stores and banks changed their business hours. Some Washingtonians missed their commuter trains, and many couldn't decide if they should have lunch at eleven or twelve o'clock, but overall reaction was positive. "Maledictions, growls, and even cusses at 7 A.M., gave way to smiles and gayety" reported the *Washington Post*, "when toils were ended at 3:30 or thereabouts, and thousands took to recreation in the sunlight." A local clothing company advertised a new reason to buy its golf suits: "Early to bed and early to rise, and you'll take up golf now, if you're wise."

FROM THE AVENUE AT NINTH

Make the Most of Your Extra Hour

Early to bed and early to rise, and you'll take up golf now, if you're wise. Of course, we carry everything that the ancient game requires in the way of wearables and accessories, priced as follows:

4-piece Golf Suits	$47.50
3-piece Golf Suits	40.00
2-piece Golf Suits	35.00
Separate Sport Coats	20.00
Separate Knickers	5.00
Golf Shoes	7.50
Golf Hose—Special values	2.85
Golf Shirts	3.50
Caps and Hats	2.00
Golf Jackets	6.50

Parker-Bridget Co.

The Avenue at Ninth
Daily: 8 to 5:30.

Taking advantage of Harding's "Ragtime."

President Harding did his part as well, rescheduling his cabinet meetings from 11 to 10 A.M., but Congress, still wary of the wrath of farmers, would have nothing to do with Harding's plan. The Supreme Court also ignored the new hours and continued to convene at high noon. The voluntary nature of the plan for non-governmental institutions meant that, while many businesses adopted the new hours, many others did not, and Harding's experiment quickly descended into widespread inconvenience and confusion.

In response, Washingtonians began using derisive names for Harding's plan, calling it "Ragtime," "Daylight Slaving Time," and "Near Daylight Saving Time"—a reference to "near beer," the not-very-satisfying low-alcohol beer substitute sold during Prohibition, which had begun two years earlier. Opponents claimed that under "Ragtime," virtually every activity, from rising in the morning to bedtime, required perpetual readjustment of one's daily habits, including the "great moral courage needed to go to bed an hour early." Moreover, because unlike DST this plan was not universally applied, many problems were caused by its uneven observance. Theaters, for instance, continued on their usual schedule, so many a theatergoer had to miss part of the final act of a play in order to catch the last train to the suburbs. Capital bankers protested that moving work hours up without changing the clock did not allow sufficient time to reconcile payments between banks.

The issue became extremely polarizing and even led to a small piece of history. On May 23, 1922, on Washington's WJH radio station, two debaters presented their views on the proposition "Daylight Saving Time is an Advantage," with the audience acting as judge. It was the first debate in the history of radio.

As the summer went on, it became clear that Ragtime was pleasing very few people very little of the time. President

Harding was infuriated by the uproar but was finally forced to capitulate. He declared that the plan, originally scheduled to run to October, would be terminated in early September and then "put aside, not only for this year, but for all time so far as this Administration is concerned." When Ragtime ended on September 4, 1922, the vast majority of federal and District of Columbia employees were happy to return to their original working hours. Ralph Norton, secretary to the police commissioner, quipped, "For once I did not have to eat my breakfast with one hand and dress with the other." After Harding's experiment, Washington would go without advanced time for more than twenty years.

By 1923, DST was in effect in a wide scattering of localities throughout the East and Midwest, but mountain and western cities such as Denver (which had tried DST for two years), Los Angeles, San Francisco, and Seattle remained on standard time. Agricultural groups continued to fight local DST laws. Representing the regional point of view, the *Houston Post* called the daylight saving scheme "a fad" and congratulated Texas for its sagacity in shunning new DST measures. "The advocates of the plan are frankly willing to ignore the opposition of the farmers, . . . the working people who must get up before daylight in the summer, . . . the women who . . . must arise earlier to prepare breakfast for their families, and all others who find the new plan a hardship."

AN HOUR IN COURT

A number of cities and states attempted to rein in the chaos by making new laws mandating the use of either standard time or daylight saving time, and establishing fines and penalties for those who ignored the law. In some areas where standard time was the official time, many people "bootlegged" daylight saving time, leading the *New York Times* to report that daylight saving time went into effect three ways: "officially, unofficially, and anti-officially." For example, both Nashua and Manchester, New Hampshire, officially kept standard time to avoid a $500 state fine, but many businesses advanced work time one hour to keep in step with Boston and other New England municipalities. Wisconsin identified daylight saving time as a public nuisance and threatened to impose fines of up to $500 and jail sentences for violators, but DST was blatantly used throughout Milwaukee and the state did little to stop it.

As new DST regulations made headway, some local laws were challenged in courts. In 1921, a theater owner, G. W. Smith, disputed the legality of Pittsburgh's DST ordinance, claiming it cost him money because customers thought it was too late to attend movies at 9 or 9:30 P.M., when it got dark. Judge John Shafer refused to grant an injunction, however, ruling that the DST ordinance was optional rather than mandatory. In 1923, DST opponents in Pennsylvania passed a state law declaring Eastern Standard Time "the sole and uniform legal standard of time," but with the prodding of Robert Garland, still a Pittsburgh councilman, the mayor of Pittsburgh continued to proclaim DST for the city each year. However, the mayor made sure to order that all public clocks display standard time, so that "officially" the city was obeying the state law.

Also in 1923, Connecticut found itself embroiled in its own legal disputes. Daylight saving time was observed in all or most parts of its three bordering states—Massachusetts, Rhode Island, and New York. But in Connecticut itself, although DST was in use in a few areas, there was strong official opposition. In May 1923 the state House of Representatives passed a bill forbidding public display of any time other than standard time, under

Led by Robert Garland, a chorus of diverse interests supported DST in Pittsburgh.

penalty of a fine up to $100. When the state Senate considered the bill, Senator John Trumble offered an amendment to permit display of DST provided the clock was labeled "Daylight Time." This prompted a DST opponent, Senator Wallace Pierson, to mockingly suggest a similar amendment—he thought such clocks should be labeled "Incorrect Time." In the end, the anti-DST bill was quickly enacted.

While most Connecticut residents complied with the new

law, some stalwart DST supporters stood firm. George Savage, holding that he preferred to display no time rather than "misleading or confusing time," ordered the clock on the tower over his company's office building in Meriden to be stopped and not display standard time until the day DST would have ended in the fall. Some ridiculed the law, such as one Connecticut resident who decried the shocking defiance of the law by Yale University's Astronomy Department, which was known to be displaying criminal clocks set to Greenwich Mean Time. A Hartford jeweler, Merton Bassett, decided to test the constitutionality of the new law. He set the clock in front of his Main Street store to daylight saving time and was arrested and fined $5. Bassett appealed to the Connecticut Supreme Court. He claimed the law restricted him in the use of his private property—his clock—and therefore represented an excessive exercise of the police power of the state. But on March 1, 1924, the court upheld the law as a reasonable legislative act with the purpose of preventing "inconvenience and confusion." Bassett paid his $5 fine.

Although only standard time was legal in Connecticut, several localities used DST illicitly. But in 1927 the citizens of the village of Hazardville stepped well beyond state regulations by utilizing *three* time systems: The churches and stores in Hazardville followed daylight saving time, and farmers in the outlying districts generally stuck to standard time. In addition, Hazardville came up with a third time, dubbed "half-time"—one-half hour advanced. The town's mills adopted half-time as a compromise, since the mills dealt with both farmers and villagers, and then Hazardville had three times in simultaneous operation. One observer commented that those who attempted to keep themselves in conformity with the three different Hazardville times had little time for anything else.

In November 1924 DST opponents in Massachusetts put

the following question on the state ballot: "Shall Daylight Saving be retained by law in Massachusetts?" The final vote was 492,239 in favor and 426,757 against—daylight saving time won, but narrowly, with only 53.6 percent of the vote. Then, in 1925, the same DST opponents tried to strike down the state law in federal court on the grounds that it conflicted with the federal Standard Time Act and was therefore unconstitutional. Several plaintiffs joined forces. The Massachusetts State Grange represented farmer objections to DST. The rural town of Hadley was concerned that its children would not get to school in time to be counted for calculating state educational aid. And there were two private citizens: a Mr. Mann, who owned farmland on both sides of the Massachusetts–New Hampshire border and alleged that the statute made it more costly for him to employ labor at the beginning of the day; and a Mrs. Snow, who complained that wives of railroad workers had to keep two standards of time, as her husband slept and ate by railroad time but her children were regulated by their school's daylight saving time.

A three-judge panel in federal district court ruled that because the U.S. Standard Time Act applied only to interstate commerce and federal operations it did not preclude a state's setting its own local time. The plaintiffs appealed their case, *Massachusetts State Grange v. Benton* (Jay Benton was the attorney general of Massachusetts), all the way to the United States Supreme Court. Before that court on October 23, 1926, the plaintiffs' attorney, Frank Morrison, argued that under the Standard Time Act, standard time was the national standard for measuring time and it was the duty of the Massachusetts attorney general to enforce it. Attorney General Benton presented the state's argument: The federal law was restricted in scope and the state law was not in conflict with it.

One month later, on November 23, 1926, Justice Oliver Wendell Holmes delivered the Court's opinion. He noted the federal court's decision—that the federal and state laws applied to distinctly different matters—and declared, "We see no sufficient reason for differing from it upon that point." The Supreme Court upheld the lower court's ruling and declared the Massachusetts DST law constitutional.

Beyond the legality of DST measures themselves, the wave of new local DST laws in the 1920s impacted the legal system in a variety of unexpected ways. In a lawsuit brought in Bridgeton, New Jersey, two brothers were summoned to appear in court at 10 A.M. When they arrived at 10 A.M. standard time, they found that under DST they were an hour late and judgment had already been entered against them. The brothers appealed to the New Jersey Supreme Court, which reversed the decision, holding that a city had no power to change court hours, which were by statute based on standard time.

In Manhattan, Nellie Kenefick was handed a summons just after midnight, on a Sunday morning, which she claimed was improper because the serving of a summons on a Sunday was prohibited by law. The server of the summons maintained that he had acted within the law because his work was based on standard time and actually took place at 11:15 P.M. on Saturday night. The dispute went to the New York Supreme Court, which ruled that on the date in question, daylight saving time was the "standard time" in New York City. The summons was set aside.

And finally, two of the least reputable opponents of DST in its history must have been William Bell and Jacob Rosenwasser, convicted murderers on Death Row at Sing Sing Prison in Ossining, New York. Bell had been convicted of killing a Long Island man and Rosenwasser was condemned to death for committing a murder in Manhattan. When they awoke on Sunday, April 30, 1922, nine days before their date with the electric chair, they realized the clocks had been advanced, effectively shortening their lives by an hour. Unless he got a stay of execution, Bell protested, the local daylight saving law had unfairly shortened his remaining life. "I don't like to lose an hour," he lamented. Nonetheless, their objections were of no avail and the hour of their executions remained unchanged.

TWO-TIMED AND SNOWED UNDER

As the 1930s began and the Great Depression descended on the industrial countries of the world, the future of U.S. daylight saving time was still up in the air. The courts had ruled that local DST laws were legal, and each American locality was left to its own devices. Some areas followed permanent DST rules, some passed a new ordinance each year, and some used DST intermittently. The starting and ending dates for the daylight saving period varied from city to city and often from year to year in the same city. The most popular DST period was five months, from the last Sunday in April to the last Sunday in September, two months shorter than the seven-month period used during the war. Daylight saving time continued to be most popular in the North and East, but the utilization of DST gradually but steadily expanded, without any large-scale organized effort. The National Daylight Saving Association had become quiescent, its founder, Marcus Marks, explaining that "daylight saving is now able to speak for itself and to stand or fall on its own merits."

In 1932, Raleigh, the capital city of North Carolina and one of the few southern cities to consider a DST measure, enjoyed a rather frantic experience with daylight saving time. In late April, the Junior Chamber of Commerce proposed that the city adopt daylight saving time, and Raleigh's city commissioners held a public hearing to consider the proposal. DST supporters attended in force, while only a few opponents to the plan voiced their views. The chief argument in favor was that DST would provide more time for recreation. Also, one speaker noted, advanced time would move the popular radio program *Amos 'n' Andy* to a more convenient hour. The Junior Chamber presented several petitions in favor of DST from merchants, bankers, and other businessmen, and the commissioners unanimously voted to institute DST in Raleigh from May 1 through September 1.

Raleigh's daylight saving time began on Sunday, May 1, only two days after the hearing, making Raleigh the only locality south of Baltimore to use advanced time. The new time was put into effect by the city government and by most businesses and schools, but almost immediately the measure ran afoul of the largest single employer in Raleigh, the state government of North Carolina. The state government's offices in Raleigh rejected DST outright, taking the position that "the State departments are not operated for the benefit of the City of Raleigh but for the people of North Carolina," all the rest of whom followed standard time. Federal government offices remained on standard time, as did colleges, hotels, trains, airlines, and other enterprises catering largely to people from out of town. The *Raleigh News and Observer* expressed a growing sentiment when it declared that the city had been "two-timed" by the city commissioners.

Confusion mounted, as some Raleighites set their clocks forward while others retained the status quo. Chief of Police Clarence Barbour took no chances and carried two watches, one set on daylight saving time, one on standard. In addition, vociferous protests were raised by those who opposed any changes to the standard time, especially farmers from the more rural areas of Raleigh.

Raleigh People Divided On Time; Many on Fence

State and Federal Departments Flatly Refuse To Recognize Daylight Time

OBSERVE STANDARD TIME TODAY IN ALL CHURCHES

Future Action to Be Decided Tomorrow; Street Car Schedules Modified

With State and Federal department menta flatly refusing to recognize the daylight saving time, effective at midnight tonight under an order adopted Friday by the city commissioners, Raleigh will today enter an era of divided time, many citizens having announced their intentions to move their watches up an hour, while many others declared that they will not do so and still others are on the fence, waiting to see what course is taken by most of their neighbors.

Eastern Standard

In order to avoid adding to the confusion caused by two systems of time, All reference to time in the News and Observer, unless otherwise noted, will be according to Eastern Standard Time.

Announcements of the time of meetings and other gatherings will specify the hour as indicated by Eastern Standard Time. Raleigh Daylight Saving Time is one hour later than Eastern Standard Time and those operating upon the time system adopted by the commissioners the city of Raleigh will be enabled to make their own calculations.

Tuesday with the course beyond that date undetermined. It was also impossible to learn what course will be followed at St. Mary's

Two-timed in Raleigh.

The protests put increasing pressure on the city commissioners, who quickly ordered another public hearing, to be held Wednesday, May 4. This time, legions of opponents of DST descended on city hall. Although the Junior Chamber of Commerce held firm in its support of DST, the preponderance of speakers now called for its repeal. The commissioners ended the hearing abruptly with the statement that they had heard enough. By another unanimous vote, they switched Raleigh back to standard time, effective that very night at midnight, only four days after Raleigh's experience with daylight saving time had begun. It was said that the experiment had at least one good effect: For those four days, Raleigh and most of the rest of the state of North Carolina had forgotten about the Great Depression.

In California, numerous attempts had been made to pass a daylight saving measure throughout the twenties. In 1930, thousands of Californians signed a petition putting the question of daylight saving time on the November ballot. The *Los Angeles Examiner* endorsed the proposal, casting its foes as the "sad-eyed and long-faced opponents of all human pleasures" and waxing enthusiastic about "the opportunity to tell the world to come to sunny California, where the sun starts the clocks and where the clocks are not set to set back the sun." The *Los Angeles Times* disagreed, predicting that DST would lead to endless confusion and "especially upset . . . the State's vast agricultural interests."

Organized opposition to the initiative was led by the motion picture industry, which argued that more daylight would decrease movie theater attendance. It would also make drive-in theaters' screening times later, decreasing their patronage as well. Harold Franklin, president of Fox West Coast Theaters, warned movie industry moguls, "Since Daylight Saving has unlimited possibilities for evil to us, we must be prepared to take our coats off and get to work against it. Statistics show that wherever Daylight Saving is in effect, it cuts theatre receipts from ten to thirty percent right off the gross." On November 4, 1930, Californians went to the polls and defeated the proposal by almost three to one. The headline of the next day's *Los Angeles Times* read DAYLIGHT SAVING LAW . . . SNOWED UNDER IN CALIFORNIA.

Despite this setback, by 1932 the whole or a portion of

fifteen states, primarily in the Northeast, observed daylight saving time. It was estimated that each spring when the thirty-five million people covered by DST sprang forward, they collectively lost 3,992 years and 255 days.

THE GOOD NEW SUMMER TIME

By the late 1930s, with about one-quarter of the nation's population living in areas that observed DST, the *New York Times* applauded "the sensible way in which daylight saving has been managed. There has been no attempt to force it down the throats of people who don't like it. . . . It has been recognized that this is a big country with a variety of interests and tastes and—if you will—prejudices." Though the policy continued to yield energy savings—for example, in 1937 the Buffalo and Niagara Electric Company estimated that it lost the sale of sixty thousand electric light hours on each day of daylight saving time—such efficiencies were no longer perceived to be its primary benefit. Instead, DST proponents concentrated on the pleasure of having an extra hour of late afternoon sunshine, a happy prospect sometimes even extolled in song.

On October 19, 1927, legendary pianist James P. Johnson and six other jazz luminaries from the Duke Ellington Orchestra, calling themselves the Gulf Coast Seven, filed into a Columbia Records studio in New York City. The recording session had been initiated by blues pioneer Perry "Mule" Bradford. All the clocks in the studio were set to standard time, as it was almost a month after New York City's DST period had ended, but that didn't stop Bradford from leading the Seven as they recorded his newest composition: "The Daylight Savin' Blues." Three years later, composer Richard Whiting and lyricist Leo Robin, fresh from a year of writing hit songs for Paramount Pictures, joined efforts to compose "When It's Daylight-Saving

Time in Oshkosh," which was soon recorded by Ted Weems and his Orchestra on the Victor Special label. That same year, two well-known songwriters, Cliff Friend and Jimmie Monaco, tapped the feelings of those opposed to DST in their own composition, "You Can Have It—I Don't Want It—Daylight Saving Time." And in 1932, singer-poet Homer Balmy composed and sang the following "croon":

THE GOOD NEW SUMMER TIME

I rise at six and call it seven,
The luncheon whistle blows at eleven,
The dinner music gets in tune
Along about mid-afternoon.

At four o'clock I start for Newry,
For half an hour I drive like fury,
Arriving there at half-past three,
Which is a great surprise to me.

The sun, though advertised to set,
Remains above the steeples yet.
Long after curfew still I hear
The sweltering bleachers howl and cheer.

I had a nightmare. Waking then,
I saw the hour was hardly ten
And twilight scarce begun to fall.
'Twas but a daymare, after all.

But despite DST's growing popularity, small problems persisted. The change to and from daylight saving time always meant an extra burden for the telephone company because of a sharp increase in the number of calls made by customers to determine the correct time. For example, on Sunday, September 30, 1928, the New York Telephone Company handled thirty-five thousand "time information" calls as compared to sixteen thousand the previous Sunday. On the other hand, stores selling sporting goods were among the first to feel the beneficial effects of the time change, as the demand for outdoor sporting equipment increased as soon as advanced time was inaugurated each spring.

The irregular observance of advanced time continued to cause disruptions for travelers. In May 1923, the Cunard Steamship Line's *Aquitania* was set to sail from New York to Cherbourg, France, just two days after New York's switch to daylight saving time. A dozen passengers missed the sailing time, but Cunard officials, having foreseen that the time change might be confusing to passengers, delayed the ship's departure, and the tardy passengers who hadn't sprung forward were still able to sail. Less fortunate was Sidmon McHie, a wealthy stockbroker, who in May of the following year thought the sailing time of the White Star Line luxury liner *Olympic* was eleven o'clock standard time rather than eleven o'clock DST. While Mrs. McHie waited at the dock, their numerous trunks already taken on board in plenty of time for the trip to Southampton, England, the hour of departure came and went and her husband, who had the tickets and passports, was nowhere to be seen. Mr. McHie arrived at the dock twenty minutes after the ship had sailed, taking with it the couple's trunks and their empty cabin, which had cost the McHies the equivalent of over $14,000 today.

The erratic progress of DST also had an impact on communications. For instance, network radio programs were affected since they were broadcast on a schedule based on the local time of the originating city. In 1926, New York City's WEAF had a network of stations in seventeen cities, and when New York switched to DST, listeners in eight network cities that remained on standard time, such as Minneapolis and St. Louis, suddenly received all WEAF programs one hour earlier than usual. In addition, during the 1920s and 1930s it was common for a radio frequency to be shared by two or more radio stations, often from different cities, each broadcasting during specified hours of the day. When DST went into effect in one city and not another, conflicts between such stations escalated. This forced the Federal Radio Commission to issue new regulations stipulating that if any of the broadcast stations sharing a frequency stayed on standard time, then all stations sharing the frequency must broadcast using standard time.

Minor problems continued to occur. In 1937, Herbert Horn, the pianist for an Italian ballet company, arrived two hours late for a performance in Chicago and claimed he was a victim of daylight saving time. His excuse? It was the first day of the daylight saving time period, and he had stopped at the home of friends before going to the ballet hall. Unfortunately for him, his friends had set their clocks one hour back instead of springing forward with the rest of Chicago. This put him two hours behind the rest of the ballet company, and the performance had to go on without its pianist.

As the use of automobiles became more common, the effect on motor vehicle accidents of shifting the daylight became apparent, since dusk and darkness were the most dangerous times for motorists. In Chicago, the local automobile association found that in October 1936, under DST, there were six fewer deaths and forty-six fewer injuries during twilight than the average number of such accidents that occurred in the previous four Octobers when standard time had been in effect. Also in Chicago that year, Judge Francis Borrelli ruled that the local law stipulating a 1 A.M. closing time for saloons referred to standard time. This meant bars could stay open until 2 A.M. during daylight saving time—giving Chicago's saloon set an extra hour of drinking all summer long.

In September 1938, daylight saving time became a useful emergency tool when a massive hurricane swept through the northeastern United States, leaving in its wake a swath of destruction and 471 deaths. Governor Wilbur Cross of Connecticut called the storm "the worst in our history." In its aftermath, chaos reigned in some areas, with roving bands of looters carting off van-loads of valuables; Governor Charles Hurley of Massachusetts had to call in the National Guard to stop looting on Cape Cod and issued the order to "shoot to kill." When September 25 approached, the date the daylight saving time period ended in Connecticut, Maine, Massachusetts, and Rhode Island, each of those states' governors decided to extend DST another week as an emergency measure, allowing hurricane-stricken communities in four states to have more hours of daylight for relief and repair.

Despite its slowly growing popularity, resistance to DST continued. In Ithaca, New York, Cornell University moved its clocks ahead one hour on May 1, 1928, in accordance with the university's policy of observing daylight saving time. All six thousand members of the university community regulated their activities under advanced time from that date until Thanksgiving. The town of Ithaca, however, decided to remain on standard time, even though much of the town's activities were related to the university. For seven months, the "town" was officially one hour behind the "gown." And local autonomy relating to daylight saving time may have reached a bizarre sort of pinnacle in the summer of 1939, in the small village of Forestport, New York. Two of the four village stores opened on daylight saving time but closed on standard time, and the other two stores stuck to standard time. The village's two hotels operated on DST, but the restaurant remained on standard time. To complicate the situation even further, both the Catholic and Episcopal churches held their Sunday services on standard time but ran their daily summer schools on DST.

THE WORLD AT PEACE

Elsewhere in the world, each country that had adopted DST as a wartime measure reached its own decision as to whether to preserve it once World War I ended. Germany, which had been the first country to adopt DST, abandoned *Sommerzeit* immediately after the war, as the German Home Office in Berlin had received a large number of petitions against it. France continued with *l'heure d'été* into the post-war period. It was estimated that in 1921 the city of Paris alone had saved 200,000 tons of fuel worth 100 million francs by utilizing daylight saving time — in keeping with Ben Franklin's prediction. But that urban benefit was not enough to sway France's rural opponents of DST, and in 1922, by a narrow 265 to 260 margin, the Chamber of Deputies voted to repeal France's DST law.

In 1923, after failing in repeated attempts to pass a new DST law, the French Cabinet decided to call for everyone and everything in France to start and stop a half hour earlier between April 28 and November 3. Skeptics immediately declared that noon trains might be rescheduled to leave at 11:30 (and thus, they quipped, be a half hour less late than usual) and theatrical performances might start at eight rather than eight-thirty, but could the government convince roosters to crow thirty minutes earlier? Then Premier Raymond Poincaré made a strong appeal

for DST based on the need to conserve coal. After some political maneuvering, DST was narrowly reinstated in May 1923.
The French continued summer time each year throughout the
interwar period.

DAYLIGHT SAVING TIME SNAPSHOT: BETWEEN THE WARS	
Africa	**North America**
Algeria	British Honduras
Sierra Leone	Canada
Tangier	Cuba
	Mexico
Asia	Newfoundland
Sarawak	United States
Europe	**Oceania**
Belgium	New Zealand
France	
Greece	**South America**
Irish Free State	Argentina
Luxembourg	Bolivia
Monaco	Brazil
Netherlands	Chile
Portugal	Uruguay
Romania	
Spain	
United Kingdom	

DST during the years 1925 to 1935.

Various other nations experimented with shifting time between
the wars. Greece initiated DST in 1932 for just two months, while
in Spain it was swiftly abolished by republicans, who perhaps
viewed it as an institution of the monarchy. On the other hand, the
Soviet Union advanced its clocks one hour in 1930 and never set
them back, going on permanent, year-round daylight saving time.

Canada had observed nationwide daylight saving time in 1918, but the following year, after the Great War ended, the Canadian House of Commons defeated a national daylight saving bill by more than two to one. As in the United States, DST was adopted locally by a number of cities and provinces, and in some places it incited clashes between modernists and traditionalists. In Quebec City, for example, there was a continuing battle over DST between "modern" city leaders, who pushed for DST, and the majority of churches in the city, who vigorously objected, preferring *l'heure de Dieu*, "God's time."

Meanwhile in the country of William Willett's birth, the annual use of DST was firmly established. In 1922 the British Board of Education surveyed local educational authorities about the effect of daylight saving time on schoolchildren and identified three positive results: morning classes were held during the cooler part of the day; children had more opportunity for outdoor life in the evening; and children could see more of their parents in the evening. The strongest objection to DST was that since children were reluctant to sleep when it was light outside, large numbers of them lost "a valuable hour of sleep." Overall, twice the number of authorities favored the Summer Time Act as considered it detrimental.

It is interesting to note that the general reluctance of children to go to sleep when it is light outside was captured years earlier by Robert Louis Stevenson in A *Child's Garden of Verses*:

BED IN SUMMER

> In winter I get up at night
> And dress by yellow candlelight.
> In summer, quite the other way,
> I have to go to bed by day.

I have to go to bed and see
The birds still hopping on the tree,
Or hear the grown-up people's feet
Still going past me in the street.

And does it not seem hard to you,
When all the sky is clear and blue,
And I should like so much to play,
To have to go to bed by day?

Parliament renewed DST yearly until finally, eighteen years after Willett's first DST proposal, the Summer Time Act of 1925 made daylight saving time permanent. It called for an advanced time period of five and a half months, from the night following the third Saturday in April to the night following the first Saturday in October.

In 1927, with daylight saving time now permanent in Britain and adopted in many other parts of the world, a memorial was erected in honor of William Willett that still stands today. It is, almost as Winston Churchill and Sir Henry Norman had visualized, a stone pillar of rough-hewn gray granite. The monument sits in Petts Wood, near Willett's Chislehurst home and the churchyard where he is buried, amid a beautiful forty-five-acre forest of fir, oak, and silver birch, now named the Willett Memorial Wood—a wood he traversed many times on his morning horseback rides. On the monument's south face a sundial shows Willett's beloved summer time, one hour in advance of the time shown on an ordinary sundial. The inscription reads HORAS NON NUMERO NISI ÆSTIVAS ("I mark only the summer hours"). The phrase is a variant on the famous inscription on a sundial near Venice, HORAS NON NUMERO NISI SER-ENAS ("I mark only the sunny hours"). The inscription on the opposite side of the monument declares that the monument and the wood surrounding it were purchased by public donations "as a tribute to the memory of William Willett, the untiring advocate of Summer Time."

Willett remembered in Petts Woods.

There are a few other tributes to Willett. A wax figure of him stood for some time among the celebrities at Madame Tussard's famous wax museum in London, and then in 1933 a new inn that opened in the Petts Woods area, outside of Chislehurst, was named the Daylight Inn in his honor.

And in 1934, Winston Churchill summed up the feelings of many who, almost twenty years after Willett's death, warmly remembered him: "Let us, then, as we put forward our clocks for another summer, drink a silent toast to the memory of William Willett, who spared neither labour nor money over a

Daylight Inn in Chislehurst.

long period of his life in his advocacy of this great reform. He did not live to see success crown his unselfish efforts. . . . But he has the monument he would have wished in the thousands of playing-fields crowded with eager young people every fine evening throughout the summer and one of the finest epitaphs that any man could win: He gave more light to his countrymen."

Britain remained on summer time throughout the rest of the interval between the wars, and DST was generally popular, though not without a few mishaps. In October 1929, Canon L. H. Evans, vicar of a parish in Eton, arrived to conduct a 6 A.M. service and found the church deserted. "Perhaps the attendant has overslept," he thought and proceeded to ring the bell himself to summon the congregation. When no one appeared at the church after he had rung the bell for some time, the vicar suddenly realized that when daylight saving time ended the night before he had not turned back his clock. Having awakened the entire town of Eton at 5 A.M., he went back to sleep.

In France, however, the confusion caused by the seasonal shift in time proved on one occasion to have far more deadly consequences. On Saturday, October 3, 1925, the Paris-Strasbourg express left Paris at 5 P.M., and steamed toward Strasbourg, the largest city in France's Alsace region. The train crew knew that France's daylight saving period ended at midnight, so when midnight came as the train neared Alsace, all train personnel set their watches back one hour. The time became 11 P.M., but the train was at the location scheduled for midnight. Thus, far ahead of schedule, the crew dutifully stopped the train and waited until the clock caught up to them and they were back on schedule again. Then they continued on toward Strasbourg.

Far ahead on the same track, a freight train was passing through Hochfelden, a town not far from Strasbourg. As midnight passed, its crew mistakenly continued to operate on daylight time and did not halt to transition to standard time. The freight train now reached each location an hour before it should have. As the freight train was switching from one track to another, the Strasbourg express, running, correctly, on a standard-time schedule, sped toward it at more than sixty miles per hour. The collision was violent and flung the express's engine and many freight cars forcefully off the tracks. The express's engineer and the brakeman of the freight train were killed instantly, and several passengers and crew members were injured.

Despite the tragedy in France, by the 1930s Willett's idea had become so entrenched around the world that the League of Nations, noting the nonuniformity of the observance of DST, recommended that all governments adopting the "so-called 'Daylight Saving' or 'Summer hour'" agree on identical dates for the beginning and the end of the summer-time period. This would allow world transportation systems to make allowance for the time change when they drew up timetables. The League's proposal was never put into effect, however, and still hasn't been today.

So although "Willett Time" had stretched across the globe, it proved extremely difficult, both internationally and within individual countries, to reach any consensus on its application, a state of affairs that continued for years—until once again the world was plunged into war.

Chapter Six
War Time

This was their finest hour.
—Winston Churchill

At daybreak on September 1, 1939, German forces launched a sudden massive land and air attack on Poland. Britain and France had stood by as Nazi Germany annexed Austria in 1938 and seized control of Czechoslovakia in March 1939. Now, believing the time had come to act, they issued Chancellor Adolf Hitler an ultimatum: Withdraw from Poland immediately or face military action. When Hitler refused, Britain and France declared war on Germany.

The Second World War would spread across the globe to embroil every major nation and the great majority of smaller countries in a war for global domination. For the next six years, the Allied forces, including the United Kingdom, France, the Soviet Union, China, the United States, and over twenty other nations, battled against the Axis Powers—Germany, Italy, Japan, and their allies—in a war fought on an unprecedented scale. By the time the conflict was over, it would cause the deaths of more than fifty million people—2 percent of the world's population.

Like the First World War, the new conflict brought renewed pressure in the combatant countries to undertake measures that would assist in the war effort. New ideas were evaluated, and old concepts were reconsidered—including daylight saving time. All of the compelling arguments for DST that had been made during World War I once again applied, especially the important contribution DST made to energy conservation.

DOUBLE IN THE SUMMER

Germany and Britain were once again on opposite sides of a world war, and once again both adopted a form of DST. This time, however, the British did not wait for Germany to act.

Moving almost immediately, just one month after the beginning of the war, Britain extended its current summer-time period by about six weeks, to mid-November. Soon thereafter the home secretary, Sir John Anderson, was pressured to move the start of summer time for the following year up to mid-February. He initially refused, citing farmers' objections, especially from Scotland, where the sun rises very late in winter, but he eventually acquiesced and pushed up the beginning of summer time to February 25.

Germany, which had not utilized daylight saving time between wars, reinstituted *Sommerzeit* for the first summer of the war, starting on April 1, 1940, and would observe DST each summer for the rest of the war. But Britain went one step further. In October 1940, the new home secretary, Herbert Morrison, extended summer time throughout the following winter and continuously thereafter, giving the British year-round daylight saving time for the duration of the war.

But it did not end there. Beginning in the spring of 1941, in an effort to improve war production, the British moved their clocks forward *another* hour in the summers. "Double summer time" was utilized for about three months of 1941, starting in early May, and for increasingly longer intervals each year of the war; by 1944, the double DST period stretched for five and half months, from April through mid-September.

Each of Britain's three extensions of summer time—first to an extended summer time, then to year-round summer time, and then to periods of double summer time—was generally well accepted. The government acknowledged the potential difficulties for farmers but considered these to be outweighed by the advantages of extra daylight for loading ships and trains and for work in factories. As the war continued and evening bombing raids by the Germans became more common, the extra light in the evening had the added advantage of allowing workers to get home safely before darkness fell.

The war exerted pressure on countries all over the world to adopt DST. Some were forced to follow it when they were occupied (primarily by Germany), but regardless of the manner of its institution, DST was a reality from Argentina to India to New Zealand.

In North America, Newfoundland, a British colony which was still separate from the dominion of Canada at the time, adopted Britain's summer-time and double-summer-time plan—but not without some bewilderment. On a summer trip to Newfoundland with two other MPs, the well-known novelist, playwright, humorist, poet, and member of Parliament A. P. Herbert reported, "Newfoundland, gallantly following the Motherland, had popped on two hours of Summer Time." But there was resistance. "Outside St. John's, the capital, the people . . . said the second hour was put on merely because the citizens of St. John's did not like driving their cars in the blackout—and who were the citizens of St. John's anyway?" Consequently, many Newfoundlanders advanced their clocks just one hour. "So in that one small place," Herbert continued, "we had to deal with two different times—Government Time and what I may call Local Rebel Time. . . . We never entered a new place or made a date, without inquiring 'What time do you use?'"

The consequences of this scheme were soon made particularly apparent to Herbert: "We arrived punctually [for a meeting] at one o'clock Local Time, two o'clock Newfoundland Government Time, 1730 British Summer Time and 1530 Greenwich Mean Time." An officer met the group and "introduced us to his Colonel. But all seemed rather awkward and vague. . . . Colonel Mackenzie, to my surprise, spoke with a strong French accent." In a short time, the group "discovered that Colonel Mackenzie had expected us one o'clock Newfoundland Government Time, twelve o'clock Rebel Time, 1630 British Summer Time and 1430 Greenwich Mean Time. We did not turn up—the Colonel assumed we were not coming and went about his duties. The French Canadian Colonel who entertained us had not the faintest idea who the three British Members of Parliament were, or why they had come." Herbert concluded, "Let this be a warning to all who try to run two times in one town."

TO PROMOTE THE NATIONAL DEFENSE

Throughout 1941, as Hitler's troops continued their offensive and Allied troops fought to stem the tide, the United States remained firmly on the sidelines. Although some urged President Franklin Roosevelt toward war, others held to the hope that the United States could avoid involvement. Just as during the early years of World War I, Europe was in the throes of war and daylight saving time was in widespread use, while the United States remained at peace. This time, however, local daylight saving time in the U.S. was observed widely: in the New England states, most of New York, New Jersey, Pennsylvania, Delaware, and in Indiana, Illinois and some parts of Virginia, West Virginia,

DAYLIGHT SAVING TIME SNAPSHOT: WORLD WAR II	
Africa	Luxembourg
Algeria	Malta
Egypt	Monaco
Morocco	Netherlands
Sierra Leone	Norway
Tunisia	Poland
Union of South Africa	Portugal
	Romania
Asia	San Marino
Ceylon	Spain
India	Sweden
Palestine	Switzerland
Sarawak	Turkey
	United Kingdom
Europe	Yugoslavia
Albania	
Austria	**North America**
Belgium	British Honduras
Bulgaria	Canada
Czechoslovakia	Cuba
Denmark	Newfoundland
Estonia	United States
Finland	
France	**Oceania**
Germany	Australia
Greece	New Zealand
Hungary	
Iceland	**South America**
Ireland	Argentina
Italy	Peru
Latvia	Uruguay
Lithuania	

DST during the years 1939 to 1945.

Florida, Kentucky, Tennessee, and the Upper Peninsula of Michigan. But still there was no national standard.

That would soon change. On July 15, 1941, with the threat of war heightening, President Roosevelt proposed a bill "to promote the national defense and the conservation of electrical energy by permitting the establishment of daylight saving time." "In these times of emergency," Roosevelt wrote to Congress, "it is . . . essential for us to insure the conservation of electricity. . . . The Government agencies primarily interested in the fullest use of electricity for national defense have advised me that there is an immediate need for the extension of this daylight saving time." The bill Roosevelt proposed gave the president authority to establish daylight saving time with a time advance of up to two hours on a regional or national basis and for such periods as he deemed necessary. Meanwhile, Roosevelt also appealed to the governors of southern states to enact DST by local statute to combat the effects of regional power shortages. In response to the president's request, North Carolina, Mississippi, and additional areas in Virginia, Kentucky, and Tennessee instituted local DST.

Congress held hearings on the president's and several other DST bills in August 1941. Some in the press commented on how the president, by seeking the authority to establish regional DST where needed rather than pushing for national DST, had succeeded in lessening opposition from farm states. Longtime DST champion Robert Garland, now seventy-eight years old, supported national DST but opposed year-round daylight saving time because of the hardship it would cause in winter. Still, as during World War I, the proposals for national DST in the United States languished as long as the country was at peace.

Then came the morning of December 7, 1941: Waves of

Japanese bombers filled the Hawaiian skies over the American naval base at Pearl Harbor, sinking or damaging almost every battleship in the American Pacific Fleet—and suddenly the United States was at war.

Galvanized by the attack, Americans immediately united in a full-scale mobilization. As vast numbers of young Americans volunteered for military service, a wide array of options were considered to support the war effort—conversion of civilian factories to the manufacture of war-related materials, the rationing of goods, price controls, internal security measures, and many more. And less than a month after the Pearl Harbor attack, proposals to utilize national DST once again as a wartime measure were introduced in Congress. The results of a Gallup Poll, released in January 1942, now showed strong public support for DST—not just in the summer, but throughout the year—with 57 percent favoring year-round DST, 30 percent opposing, and 13 percent undecided.

In the debate on the DST legislation, wartime considerations were paramount. DST would help the war effort, contended General J. L. DeWitt of the Western Defense Command, because "defense production would be accelerated, power conserved, morale raised, and safety during blackouts better insured." Though the general's purview was limited to the Pacific Coast, he emphasized that the DST plan "should be of national scope." Many in Congress favored year-round DST, instead of World War I's seven-month period, to help lower evening peak electrical power loads, which were greater in winter. Congressman Clarence Lea of California emphasized that power consumption was highest at factories and offices between 5 and 7 P.M., and if this peak load were reduced, enough electricity could be saved each year to produce over seventy million tons of much-needed aluminum for planes.

The DST bill received some predictable opposition from rural areas, but many farm-state congressmen felt they could not or should not object to DST as a war measure. "I hope that gentlemen representing agricultural sections will not persist in offering opposition to this measure upon the ground that it will inconvenience the farmers of this country," declared Representative Edward Cox of Georgia. "I am a farmer, my people are farmers. . . . There is not a more patriotic group in the world than the farmers of America. . . . If the making of sacrifices and the assumption of hardship adds to production, and therefore to strengthening national defense, you will not find any people anywhere more willing, more anxious and more enthusiastic in the assumption of . . . these new burdens, even though they impose hardships, than our farmers."

Rising to the needs of the moment, Congress passed "An Act to Promote the National Security and Defense by Establishing Daylight Saving Time." It provided for national DST but barred the president from fine-tuning the details and stipulated that the law would cease to be in effect six months after the termination of hostilities. Roosevelt accepted the limitations and signed the bill into law on January 20, 1942.

National year-round daylight saving time, unofficially known as "War Time," became effective on February 9, 1942, and superseded all local daylight saving time laws. Prior to the war, a DST-like effect had already been caused in some areas by moving time zone boundaries west. Nationwide DST now advanced the clock another hour, yielding some very late winter sunrises in these areas. Local officials appealed to the Interstate Commerce Commission to shift time zone boundaries back to reduce the dark mornings, but the ICC would not change time zones solely to nullify the effects of War Time. Even so, some of the affected areas adopted a local time that

---

77TH CONGRESS, PUBLIC LAW 77-403

AN ACT TO PROMOTE THE NATIONAL SECURITY AND DEFENSE BY ESTABLISHING DAY-
LIGHT SAVING TIME

*Be it enacted by the Senate and House of Representatives of the United
States of America in Congress assembled, That:*

Section 1. Beginning at 2 o'clock antemeridian of the twentieth day after the
date of enactment of this Act, the standard time of each zone . . . shall be
advanced one hour.

Section 2. This Act shall cease to be in effect six months after the termination of the present war or at such earlier date as the Congress shall by con-
current resolution designate, and . . . the standard time of each zone shall
be returned to the mean astronomical time of the degree of longitude governing the standard time for such zone.

Approved, January 20, 1942

Once again, DST as a wartime measure.

was one hour behind War Time—thus giving themselves year-round standard time during the war.

Although many DST opponents accepted War Time as a patriotic measure, others fought in Congress to repeal it, and some tried to take local action against it. In February 1943, State Representative X. T. Prentis and forty-six colleagues introduced a bill in the Iowa legislature to return Iowa to standard time, so as to "increase food production and win the war." A farm-versus-city fight developed. "I don't know any farmer who favors daylight saving," reported a Central City farmer, but a city official in Osceola observed, "Many people in the state are well pleased with the present time." Jumping the gun a bit, the *Des Moines Register* on March 25 ran the headline IOWA HOUSE 'REPEALS' WAR TIME. But the state Senate refused

DST saved energy for the war effort.

to go along with the measure, instead passing a resolution that urged businesses and local organizations in agricultural communities to change their hours to standard time from June through September. A survey of the state in July, however, found that not one single community had altered its business hours. The *Register* reported, "Farmers, irrespective of the time on any clocks, are working from sunrise to dark—and often after dark. Generally, farmers don't like war time and wish they could get rid of it but have made the necessary adjustments and have become accustomed to it."

Iowa repeals War Time.

In May 1944, the U.S. House of Representatives considered a bill to restore standard time nationwide. The army and navy strongly opposed such a measure, and Donald Nelson, chairman of the War Production Board, declared that "a return to standard time now would be a blow to the war effort which the country cannot afford." The bill foundered in Congress and even the leading opponents of DST refrained from pressing for action.

Victory in Europe for the Allies came on May 8, 1945, and on September 2, Japanese officials signed the documents of their surrender aboard the battleship USS *Missouri* in Tokyo Bay. World War II was over. Some American radio announcers, who for almost four years had been announcing the local time as "Eastern War Time," now delighted in saying, "It's eight o'clock, Eastern Peace Time." But the future of daylight saving time would be anything but peaceful.

THE BUCKOO THAT TWISTED THE CLOCK

When World War II ended, some countries immediately abandoned daylight saving time and others retained it. Among the nations that kept summer DST, there was a wide discrepancy in the dates selected for the period. In 1948, for example, Denmark's daylight saving period ended on August 8; Israel's on September 1; Japan's on September 12; Canada's on September 26; Czechoslovakia's, Ireland's, Newfoundland's, and Portugal's on October 3; and Great Britain's on October 31.

In occupied Germany, the Allied Control Council now set the time, and it put Germany on *Sommerzeit* in 1947, advancing the clock one hour on April 6. Postwar energy shortages soon caused it to institute double *Sommerzeit*, on May 11. But when German citizens protested that the use of double DST was causing too much hardship, the council relented and on June 29 German clocks were set back to single *Sommerzeit*. As a result, in less than three months, Germans had three one-hour time changes—standard time to DST to double DST to DST again.

Britain revoked many of its wartime laws, including year-round DST and double DST in the summer, and returned to the prewar summer time. But in 1947, a fuel crisis led the government to extend daylight saving time again. When the bill was introduced before the House of Commons, the recently knighted Sir Alan Patrick Herbert, now a vocal opponent of summer time, declared, "I agree, reluctantly, that we must have this Bill, because of the crisis. But I cannot miss the opportunity of attacking, on behalf of thousands of citizens, the principle of Summer Time. I hate to think that Big Ben, that great bell which has been such a voice in the councils of the world, will be heard booming over the B.B.C., first Berlin Time, and for the rest of the summer Moscow Time." Sir Alan viewed summer time from a patriotic perspective: "Surely this country should be the last to abandon Greenwich Time. . . . I am sure that, if Hitler had conquered the world, the first thing he would have done, having a certain imagination, would have been to say: The Prime Meridian shall run through Berlin and not through Greenwich. . . . It is one of the great glories of this country that all nations have agreed that Greenwich and Greenwich Time shall be the centre of all astronomy and navigation. . . . That is a thing we must not throw away. Let the Empire go, if you must, but cling fast to the Prime Meridian."

British agricultural interests, though stressing the difficulties double summer time would cause them, nevertheless went along with it as necessary in the emergency situation. The bill was passed and Britain employed extended summer time plus double summer time for one year. In 1948, with the fuel crisis over, Britain reverted back to regular summer time, a policy that would continue for twenty years.

In America, just three days after the Japanese surrender, several members of Congress, unwilling to wait until the DST law automatically expired six months after the war's end, introduced bills to repeal the law immediately. Congressman Clarence Cannon of Missouri declared that national daylight saving time "should be the first of the artificial war expedients to be removed," and the War Production Board formally withdrew its opposition to DST repeal, as did the army and navy.

Congressman Cannon was forceful in his complaints against War Time: "It is bringing about wastage of manpower on the farm and absenteeism in the factory, . . . it increases the consumption of power and light. . . . It has increased the number of traffic accidents on darkened highways. . . . It is sending millions of children into the darkness and cold to await school buses on remote highways and unguarded streets. . . . It has

deprived millions of an added hour of sleep on sultry nights."
Maryland Representative Dudley Roe agreed: "War Time has
made our people get up in the dark, dress in the dark, milk
their cows in the dark, send their children down the long lanes
of our farms to the public highway and get on school buses in
the dark. I rejoice at the repeal of this foolish regulation." And
Congressman Benton Jensen of Iowa added piously, "We
should abolish war time and get back to God's time."

To bolster his argument, Cannon reported that he and other
proponents of repeal had received a flood of letters and
telegrams of support from all sections of the country, both rural
and urban, and from all classes and professions. Among the
337 letters he submitted to the *Congressional Record* was this
poetic entreaty:

Dear Congressmen, Senators, President too:
> Give ear to a pitiful rhyme.
And pity the farmers that fell in the soup when the Government
> doctored the time.
I'm a man of few phrases, me learnin' is scant, and I'm longin' to
> make meself clear.
It's bad in the summer, it's worse in the spring, and it's fierce in the
> fall of the year.

In balmy Siptember I rise from me bed and I dress be the light o' the
> stars.
And finish me dreams as I wait for the cows, with me head hangin'
> over the bars.
The cock on the roost sees me lantern go by, and he thinks it the
> morning's first gleams.
But e'er he's done crowin' he's left in the dark, and the chickens all
> laugh in their dreams.

Me pigs are reposin' on pillies of mud, and me horses are sprawled in
  the stall.

And I question the wisdom that sent me abroad an hour too soon in
  the fall.

I'd willingly go to me bed with the birds and be up with the song o'
  the lark,

But curs'd be the notion of savin' the day by blundrin' about in the
  dark.

I hope when the guilty one knocks at the gate, for Peter to open the
  lock,

He'll say: "Step aside; you've an hour to wait. You're the buckoo that
  twisted the clock."

J. Ward Williams
Prospect, New York

It took only about three weeks from the war's end for Congress to pass a repeal bill and for President Harry Truman to sign it into law. The repeal act terminated federally mandated daylight saving time as of September 30, 1945. On September 29, the *Des Moines Register* gave instructions on clock changing to those Iowans unused to dealing with DST so as to "get you and your household back to standard time. It's really very simple." An extra hour of sleep was the reward, and the *Register* assured its readers, "Your favorite Sunday radio programs will come on at the same hours they did before. *The Quiz Kids* you are used to hearing at 6:30 P.M. will be heard at 6:30 P.M."

79TH CONGRESS, PUBLIC LAW 79–187

AN ACT TO PROVIDE FOR TERMINATION OF DAYLIGHT SAVING TIME

Be it enacted by the Senate and House of Representatives of the United States of America in Congress assembled, That:

Notwithstanding the provisions of the Act of January 20, 1942, . . . at 2 o'clock antemeridian on Sunday, September 30, 1945, the standard time of each zone . . . shall be returned to the mean astronomical time of the degree of longitude governing the standard time for such zone.

Approved, September 25, 1945

Postwar repeal, again.

Thus ended the era of War Time, and once again, each locality was free to make its own choices about daylight saving time, irrespective of its neighbors. But it was this freedom that was to lead to the most chaotic epoch in the history of DST.

Chapter Seven

Clock Chaos

The time is out of joint; O cursed spite
—William Shakespeare, *Hamlet*

*A*fter national daylight saving time was repealed, each state, county, and municipality in America could decide whether and how it would institute DST. In many localities, prewar daylight saving time laws came back into effect and by 1947, about fifty million Americans, roughly 40 percent of the population, were observing DST.

LOCAL OPTIONS, LOCAL PROBLEMS

Throughout this early postwar era, the president of the United States would weigh in on the matter. President Harry Truman was not opposed to national daylight saving time, but he felt that DST was unworkable when each locality made its own decision. "I don't like piecemeal daylight saving here and there over the country," he argued. At a White House press conference in 1948, Truman declared that local-option DST was "a lot of 'hooey.' . . . We built those time zones so that [there would not be] time differences of fifteen minutes between cities ten miles apart." Under local-option DST, "We now get an hour's difference in cities that adjoin each other. If there is a daylight saving program, it should be a national program, not a local one." Truman would sustain his steadfast opposition to local DST throughout his administration. Near the end of his term, in June 1952, he took an airplane flight with Mrs. Truman and recorded the following in his personal diary: "We landed at Grandview Air Port at 1:40 central standard time — 3:40 Eastern Day Light Time — a monstrosity in time keeping."

"In the good, old summertime."

Many were happy to retain DST.

But much to Truman's chagrin, scattered local use of daylight saving time continued throughout the postwar period. Ongoing resistance to DST was perhaps best captured by the Canadian writer Robertson Davies, whose title character in *The Diary of Samuel Marchbanks* expounded the opinion of many: "I object to being told that I am saving daylight when my reason tells me that I am doing nothing of the kind. I even object to the implication that I am wasting something valuable if I stay in bed after the sun has risen. As an admirer of moonlight I resent the bossy insistence of those who want to reduce my time for enjoying it. At the back of the Daylight Saving scheme I detect the bony, blue-fingered hand of Puritanism, eager to push people into bed earlier, and get them up earlier, to make them healthy, wealthy and wise in spite of themselves."

In the years following World War II, daylight saving time at the local level resulted in several exceptional circumstances. Consider, for example, the unusual situation of the desert town of Palm Springs, California. In summer, the desert sun beats down mercilessly, leaving the town's residents with little desire for an extra hour of daylight. In the winter, however, the eleven-thousand-foot summit of Mount San Jacinto casts an afternoon shadow over the town and the sun effectively disappears by 3 P.M. As a winter resort, Palm Springs was eager to capture more hours of daylight, and in 1946 its city council unanimously passed a unique ordinance establishing daylight saving time—but only for the winter months. This rare form of DST, which locals called "Palm Springs Time" or "Sun Time," was launched on November 17, 1946. At first, both tourists and locals seemed to welcome the change. But after a few weeks it was clear that the confusion caused by being out of sync with the rest of California was much more significant than the extra hour of sunlight. When a petition signed by 284 residents was presented to the city council requesting the repeal of Palm Springs Time because of its "inconvenience to the majority of the residents and local businesses," supporters, though still asserting its benefits, agreed to the repeal "in the interests of local unity." On January 12, 1947, less than two months after it had begun, the Palm Springs City Council voted unanimously to discontinue Palm Springs Time.

There were other local oddities. In 1950, the tiny central Illinois town of Minonk stood alone. The city council had set April 23 as the date to change to daylight saving time, not realizing that all of its neighboring communities, and indeed every area observing DST throughout America, would commence DST on April 30 or later. But by the time the discrepancy was realized, Minonk's weekly newspaper was already proclaiming April 23 the first day of DST and Mayor John Ketchmark felt it was too late to revise the date. So for one week, Minonk's 2,500 citizens observed DST while 150 million other Americans did not.

In the early 1950s, local railroad lines in large DST cities generally ran on daylight saving time, but long-haul railroads in the United States and Canada continued to operate under standard time—which sometimes led to dangerous situations. On one occasion in 1951, a Canadian National Railway train was steaming from Hodgson, Manitoba, to Winnipeg, carrying passengers and picking up boxcars bursting with Manitoba grain. The journey was uneventful until about noontime. "Suddenly, the engineer began sounding the whistle in short rapid blasts and simultaneously the air brakes were applied in emergency," Brakeman Allan B. Peden recalled. "As the train came to a grinding stop, the conductor . . . stuck his head out for a look-see. He yelled back to us, 'We hit a car.' We saw an automobile with both front doors open, down in the ditch on the fireman's side. Two women, standing on the road close by, were shouting hysterically about a baby in the car. The tail-end brakeman, who was a few steps ahead of us, ran over and retrieved a tiny baby, still sound asleep in its blanket, from the front seat of the car. The child, luckily, was unharmed as were both women. . . . One of the women, presumably the driver of the car and mother of the child, mentioned repeatedly in her sobbing voice that the train was due by this area an hour ago and that daylight saving time had her all confused." Peden commented, "Well, confused she certainly was because we were running right on time."

Throughout the late 1940s and early 1950s, the blooming of cherry blossoms was not the only rite of spring in Washington, D.C.; another was the debate in Congress on whether to continue DST in the nation's capital for another summer. A poll taken by the Metropolitan Police Force showed that 73 percent of Washingtonians favored DST, but senators and representatives from rural areas echoed their constituents' disdain for it. Louisiana Senator John Overton once threatened to mark the debate by inserting the following announcement in a Washington newspaper's lost-and-found column: "Lost—somewhere between sunrise and sunset, one golden hour, set with 60 diamond minutes. No reward is offered; it is lost forever." But Rhode Island's Senator J. Howard McGrath favored DST, reminding his colleagues that while the senators themselves returned to their local districts for the summer, D.C. residents would be forced to sweat out the scorching Washington summers. "[We must] open our hearts and be good-neighborly and kindly to them," McGrath implored. Each year DST was eventually approved, until finally, in 1953, Congress passed and President Dwight

D. Eisenhower signed into law a bill authorizing the capital to observe daylight saving time each year.

By 1953, daylight saving time was observed by eleven states in the Northeast and Far West and by some communities in eleven additional states. Regional opinions were reflected in a Gallup Poll taken in the mid-1950s. DST received strong support in the Northeast (82 percent in favor), moderate support in the West (57 percent), mixed feelings in the Midwest (51 percent), and general opposition in the South (39 percent).

The typical daylight saving time period was five months, May through September. In 1954 and 1955, several states and cities extended DST to the end of October. When a similar one-month extension was proposed to New York City's DST in 1955, the League of New York Theatres fought it, noting that theater business fell off 10 percent with the switch to DST in April. But M. D. Griffith of the New York Board of Trade responded by pointing out that "the success of the theater in New York or anywhere else depends not upon the sun but upon the stars," and the extension was approved.

CLOCK CONFUSION CONFOUNDS CITIES

By the late 1950s and early 1960s, a patchwork pattern had emerged throughout the country as states, counties, and municipalities established DST periods of varying lengths, all interspersed with states, counties, and municipalities with no DST at all. The situation had reached the point of "crazy-quilt confusion," as it was described by *U.S. News & World Report*.

TIME OBSERVED IN MAJOR U.S. AND CANADIAN CITIES IN 1955			
DAYLIGHT SAVING TIME		**YEAR ROUND STANDARD TIME**	
May through September	May through October		
Baltimore	Boston	Atlanta	Memphis
Cleveland	Buffalo	Birmingham	Milwaukee
Los Angeles	Chicago	Cincinnati	Minneapolis
Louisville	Hartford	Dallas	New Orleans
Montreal	New York	Denver	Omaha
St. Louis	Philadelphia	Detroit	Salt Lake City
Toronto	Pittsburgh	Houston	Seattle
Washington, D.C.	Providence	Kansas City	Portland, Oregon

Even among larger cities, DST usage was fragmented.

Over the course of a single year in Iowa there were twenty-three different combinations of start and end dates for DST in use, and that's not counting the town of Hopkinton, where banks opened on daylight saving time and closed on standard time. In parts of northern Idaho, daylight saving time was observed on a door-to-door basis, and shoppers had a difficult time knowing which stores were open when.

Des Moines Sunday Register

Des Moines, Iowa, Sunday Morning, September 6, 1964 — *Local News Section*

Daylight Time Spreads Nightmare of Chaos Across Iowa

An Iowa nightmare.

Nashville, Tennessee, operated on a system that some citizens jeeringly called "confusion time." Daylight saving time was adopted by TV stations, schools, hospitals, and many, but not all, businesses. Nashville's two newspapers were published in the same building, and each had a big clock on the front of the building. The *Tennessean*, editorially supporting DST, moved its clock's hands an hour ahead to display Central Daylight Time, but the *Banner*, which opposed DST, kept its clock set to Central Standard Time. Appropriately, Nashville county music legend and Grand Ole Opry star "Grandpa" Jones expressed his opinion of daylight saving time in song:

DAYLIGHT SAVING TIME

For years and years folks got along with an old grandfather's
    clock
Or just a common old sundial sittin' out on a block
And then the dollar watch it had its day, and the wristwatch it
    was fine
Then along came a man who ruined us all with Daylight
    Saving Time.

Well, I'm a huntin' that man who first thought up this Daylight
    Saving Time
Until he moved the clocks around ev'rything was a-goin' fine
But I guess he's hid and won't come out, he knows he's out of line
But I get up late, I have to wait,
Can't keep it straight, who did he hate
I mean the man who first thought up Daylight Saving Time.

And after charter bus operators, bewildered by daylight saving time differences, delivered hundreds of fans to a Notre

Dame football game an hour after kickoff, the band played the popular song "I Didn't Know What Time It Was."

Many communities simultaneously followed standard time for endeavors governed by federal law, such as interstate commerce, and their own local DST, when applicable, for everything else. In Pennsylvania, official state business was conducted on standard time, but DST was observed in every one of the state's six hundred communities, except for tiny Yatesville (population 472). State laws in Maine, Iowa, and Ohio required liquor stores to open and close on standard time, even in localities where everything else operated on DST. And fishermen in the parts of Montana observing DST might bait their hooks on Mountain Daylight Time, but they had better cast their lines on Mountain Standard Time, since that was the basis for state fishing licenses.

One of the more notable examples of mismatched times arose in Minnesota's Twin Cities of Minneapolis and St. Paul. The cities lie on opposite sides of the Mississippi River, with their downtown areas just a few miles apart. Minneapolis is Minnesota's largest city and St. Paul is the second largest and also the state capital. Although two distinct cities politically, together they form one metropolis many of whose residents live in one city and work or engage in recreation in the other.

Thus it was front-page news when the St. Paul city council voted in the spring of 1965 to begin daylight saving time on May 9 while the Minneapolis city council confirmed that their city would continue to follow Minnesota state law, which stipulated a DST period starting on May 23. St. Paul's mayor, George Vavoulis, declared that he had proposed the early start for DST because most residents wanted their time to conform with that of nearby Wisconsin and with most of the rest of the nation. "There has never been an issue," Vavoulis contended, "in which the will of the majority has been so clear," to which Minneapolis's mayor, Arthur Naftalin, replied that the state law had to be obeyed. When Minnesota governor Karl Rolvaag weighed in he did not mince words, condemning St. Paul's action as "rash, short-sighted, extreme, and unwarranted."

With some Minnesota state legislators backing a bill to force compliance with the state DST law, Mayors Vavoulis and Naftalin appeared before the legislature and pressed for an extension to the statewide daylight saving period, which would solve their mutual problem. But neither side was able to muster enough support, and there was a stalemate in the legislature.

Then, in an unprecedented move, Minneapolis's Mayor Naftalin appeared before the St. Paul council. He implored the councilors to rescind their action and use the state-mandated DST dates—but the council merely reaffirmed its decision to have DST start the following Sunday.

The next day, both city councils held a joint meeting at the Criterion Restaurant in St. Paul. Mayor Naftalin, his voice rising with each sentence, called on St. Paul to reconsider its "irresponsible" unilateral action, since having St. Paul on daylight time and Minneapolis on standard time "would be intolerable. . . . Enormous confusion would result." But Mayor Vavoulis instead urged the Minneapolis councilors "to reassess their position and join with St. Paul." He produced "a flock of telegrams" from Minneapolis citizens and businesses commending St. Paul's action and expressing the wish that Minneapolis would follow suit. With all sides standing firm, daylight saving time came to St. Paul at 2 A.M. on Sunday, May 9. One newspaper headline read CLOCK CONFUSION CONFOUNDS CITIES.

With the state law mandating standard time and St. Paul

St. Paul Pioneer Press

First Newspaper In Minnesota

In Sunday's Pioneer Press

VACATION-TRAVEL SECTION
See this special section designed to help you plan your vacation. You'll find it packed with resort listings, travel tips and the latest in fashion trends. See Sunday's Pioneer Press.

26 PAGES   ST. PAUL, MINN., WEDNESDAY, MAY 5, 1965   c☆   TELEPHONE 222-5011   PRICE 10 CENTS

Clocks in Minneapolis, St. Paul Will Conflict

Nonidentical twins.

going onto DST, timekeeping in St. Paul's suburbs quickly fell into disarray: Eighteen suburbs shifted to daylight time, nineteen stayed on standard time, four kept standard time but advanced the hours of local town offices, one went unofficially on daylight time, one stayed on standard time but its businesses observed daylight time, and two left the time up to each citizen's individual choice. Suburban St. Paul school districts made their own decisions on time, with two scheduling their classes based on DST, eight continuing on standard time, and three keeping on standard time but moving class times one hour earlier. A. H. Pagenkopf of the Mounds View School District summed up the situation: "We're in the middle of a muddle."

The St. Paul fire department switched to DST but the police department stayed on standard time. Two St. Paul policemen showed up for work wearing a watch on each wrist, one for standard and one for daylight time. But after just one day of utter confusion—with St. Paul police officers not knowing when they should enforce rush-hour parking regulations—the police department adopted DST. Minnesota's state capitol and other state offices, although located in St. Paul, followed the state law and remained on standard time.

The best solution may have been the one offered by Al Olsen, the St. Paul city council recorder, who told a reporter: "I don't have any watch. I'm going to lunch when I'm hungry."

Even leisure-time activities were affected. State Attorney General Robert Mattson issued an opinion that liquor stores and bars must follow the state law, regardless of local time changes. This allowed St. Paul's bars to stay open an hour later than usual. All Twin Cities television stations continued to follow standard time, and the TV listings in St. Paul newspapers showed standard time but advised citizens in DST communities, "Just remember 7 P.M. shows will be on at 8 P.M."

The chaos finally ended on Sunday, May 23, 1965, when Minneapolis and the rest of state law–abiding Minnesota shifted to daylight saving time. The Twin Cities were, at least on their clocks, in sync once again.

THE WORST TIMEKEEPER IN THE WORLD

Despite the growing popularity of DST in the postwar era, its lack of uniformity became a matter of increasing national concern. Consider that for five weeks each year, from the end of September to the end of October, Boston, New York, and Philadelphia were not on the same time as Washington, D.C., Cleveland, or Baltimore—but Chicago was. Businesses had trouble determining the hours distant businesses were open and coordinating with remote markets, while business travelers missed important appointments because they used the wrong time standard. The situation afflicted everyone from business executives to vacationers to shoppers to long-distance telephone callers, among countless others. Scheduling a doctor's appointment, a movie outing, or a dinner date across county lines, or sometimes just across town, required checking the time differences.

Broadcasters also had to deal with the inconsistent use of daylight saving time. In the early 1950s, radio networks broadcast major programs in the East on DST, while allowing

It's difficult to fly straight with uncoordinated wings.

optional rebroadcasts an hour later by stations in the standard time areas of the South and West. New facilities and equipment were needed to replay live TV programs, and by 1965 the television industry was annually spending the equivalent of over $12 million today to tape and save programs. In two years, AT&T spent the equivalent of $60 million today to expand its communication capacity for rebroadcasting.

But the time confusion was most acutely felt in the transportation industry. In 1965, buses traveling between Chicago and Minneapolis experienced five schedule changes in a six-month period, and when the timetable of a bus running through Salisbury, Maryland, was revised four times in six months, a bus company official remarked, "Our passengers were mad enough to blame us and use their own cars."

Probably the most extraordinary example of the disarray in transportation was the plight of passengers on the interstate bus following Route 2 from Steubenville, Ohio, to Moundsville, West Virginia. Traveling less than an hour and going only thirty-five miles, the bus went through seven different official time zones. If the driver and passengers had tried to keep to correct local time as they dipped in and out of daylight saving time, they would have had to reset their watches once every eight minutes.

Federal law required that long-distance railroads operate on standard time, but many of their passengers lived in DST areas. So railroads published two timetables, one in standard time and one using DST, and passengers were often baffled. Even locations that remained on standard time were not spared. Although Arizona did not observe daylight saving time, on a day when twenty other states shifted to DST, trains and buses serving Tucson had to make thirty-eight changes to their schedules. It was increasingly difficult for bus, train, and airline timetables to be printed fast enough to keep up with the time changes, and timetables that attempted to include all the local DST switchovers became so complicated that most passengers could not understand them. Travelers who missed their buses or trains because of a time shift usually blamed the bus company or railroad, and the yearly cost to the railroad industry of adjusting operating schedules and printing timetables was the equivalent of over $12 million in today's dollars.

Occasionally such glitches could be fortuitous. In September

"Sure Would Help If All Daylight Saving Time Would Start Last Sunday In April...And End Last Sunday In October!"

Uncertainty in the transportation industry.

1969, Darrell Hardesty and his wife were returning home to Linton, Indiana, from a visit with their daughter in Fayetteville, North Carolina. When they arrived at the airport, they to found that their flight home had taken off without them—they had forgotten to take daylight saving time into account. With new flight arrangements, the Hardestys arrived home in Indiana at 2 A.M., and went straight to bed. The next morning they called Mrs. Hardesty's sister to tell her of their delayed return. Only then did they discover that the connecting flight they had been scheduled to take from Cincinnati to Indianapolis had collided with a small single-engine Piper Cherokee, and all eighty-three passengers and crew had been killed. Relatives waiting for the Hardestys at Indianapolis's Weir Cook Airport had, of course, been notified of the crash and had left the airport believing the worst—until the phone call came the next morning.

Even as proponents of a national standard argued that DST uniformity would alleviate problems in communication and transportation, opposition to daylight saving time remained passionate in many areas of the country. Attendees at a meeting about DST in Knoxville, Tennessee, were so opposed to any "tampering" with local clocks that Thomas Pyne, the Interstate Commerce Commission examiner to whom all complaints about DST were funneled, had to be escorted through the crowd by U.S. marshals. At another meeting in Indiana, the beleaguered Pyne was approached by a preacher who described his dilemma. "Half my congregation favors daylight time, and the other half, standard time. When should I open my church?" Pyne answered, "Use your own judgment." "I just did that," the preacher replied, "and lost half my congregation."

In 1957, after a law prohibiting even "voluntary observance" of DST was passed in Tennessee, a Nashville insurance company rebelled and moved its clocks ahead an hour. A local judge imposed a $25 fine, and the Tennessee State Supreme Court upheld the new law on appeal, noting that "inconvenience, confusion, and conflict arise" if one business operates on a different time than its surroundings.

In 1963, *Time* magazine described the time situation in the United States as "a chaos of clocks." The *New York Times* noted the "time schizophrenia that plagues the nation" and decried the "exasperation and expense that result from the annual timekeeping pandemonium." Dr. William Markowitz, a scientist specializing in time at the United States Naval Observatory, rated the United States "the worst timekeeper in the world."

Transportation executives were at the forefront of reform efforts. In 1961, at a meeting of the Transportation Association of America (TAA), Fred Ackerman, chairman of Greyhound Bus Lines, contended that the industry needed "uniformity in time, whether standard or daylight, and uniformity in the dates of time changes within each time zone." The TAA adopted this goal as its policy. A year later the TAA and other groups plagued by daylight saving time uncertainties formed the Committee for Time Uniformity. The committee passed no judgment on the merits of DST, but instead concentrated on trying to make DST start and end dates uniform. The group steadily expanded to represent the DST concerns of many transportation, communication, finance, travel, and other interests, and focused on lobbying Congress for legislative action.

The logo of the Committee for Time Uniformity depicts a time-befuddled citizen.

As pressure grew for some sort of national standard, the Senate held a hearing in April 1963 to discuss legislation for DST uniformity. Senator Gale McGee of Wyoming, citing the problems caused by varying time standards, observed that "the clock chaos is enough to confound Father Time himself." Remembering the bitter opposition to DST from rural areas after the world wars, many senators remained wary of voicing support for any national DST legislation. But the concept of

DST uniformity—as distinct from the merits of DST itself—had gained support even from some leading farm groups. Farmers felt the pains of patchwork DST when they sold their products. Also, like most Americans, they had become more focused on television, which was generally based on the time in major cities. As farm opposition to uniform DST was diminishing, the political power of farm interests was also waning—a result of gradual population shifts from rural to urban and suburban areas.

But 1964 arrived with no federal legislation on DST uniformity. In different parts of the United States that year, daylight saving time began on April 24, 25, 26, 27, 28, and 30; May 1, 3, 14, 15, 17, 18, 24, 25, 29, 30, and 31; and June 1 and 7. It ended on August 29 and 31; September 1, 6, 7, 8, 13, 14, 21, 27, and 28; and October 4, 25, and 31.

The Commission on Intergovernmental Relations, made up of federal, state, county, and local officials, reported that the time situation was so disorganized that there was no single agency that could "provide information as to precisely what time prevails on a given date on a community-to-community basis." That year, fourteen different bills for DST uniformity were considered by congressional committees, but when Congress adjourned, none had come to a vote in either house.

And in Iowa, Miss Lillian M. Green recalled a Benjamin Franklin saying in a lyrical missive on daylight saving time that she sent to the *Des Moines Tribune*:

A KILOWATT SAVED IS A PENNY EARNED

Backward, turn backward, Oh time in your flight,
Give us more "do" light before it is night.
One daylight hour at the end of the day
Means sixty more minutes for work or for play.

Awaken Des Moines, let us follow the trend,
Set by villages, hamlets and towns without end.
If the first hour's murder to leave the old sack,
Just remember next fall we can have it all back!

UTA FOR THE USA

By 1965, with daylight saving time in force statewide in fifteen states and in parts of sixteen others, a general consensus had emerged in Congress on the need for time-uniformity legislation. Even farm-belt Iowa, the site of twenty-three different DST periods, had by now passed a resolution requesting a federal law to standardize daylight saving time throughout the country. Uniform nationwide DST also had the support of President Lyndon Johnson. Once again more than a dozen different bills were introduced in Congress. Some proposed mandatory nationwide DST, some allowed statewide exemptions, and some allowed each locality to make its own choice as long as all used the same DST transition dates.

The most popular proposal was for DST from the last Sunday in April to the last Sunday in October, which was the most commonly used period across the country. Several members of Congress, mostly from rural areas, still opposed any DST bill, and a few senators had posted signs on their doors that read THIS OFFICE OPERATES ON GOD'S TIME. Others pointed out that everybody "favors uniform time as long as everyone else will conform to the system which he favors."

But it was the concept of a nationally mandated six-month daylight saving time period—lasting until the end of October—that caused the most controversy, especially among rural legislators. "I cannot support uniformity at the cost of the many young schoolchildren who will stand cold and lonely awaiting their school bus on rural roads," asserted John

Schmidhauser of Iowa. H. R. Gross, also of Iowa, cried out dramatically: "I am not going to vote today to make myself part of a tragedy on the highways of Iowa where school children, coming across the highway to catch a school bus in the darkness and semidarkness of late fall, are mowed down by a truck or a car. . . . Let the blood be on your hands, not mine!" In the wake of these objections, Mark Andrews of North Dakota advanced the idea of a DST period from Memorial Day to Labor Day, but his proposal was defeated.

Another 1960s DST concern.

Debate continued into 1966. DST was now observed statewide in eighteen states and in parts of eighteen others; fourteen states remained on standard time all year. One hundred million Americans had an hour more daylight in the evening, while ninety million other citizens (and many cows) continued to enjoy their hour of sunlight in the morning. A few small areas of two Central time zone states, North Dakota and Texas, set their clocks back one hour unofficially, as they had for many years, so that they could be on Mountain Standard Time and have what might be called "reverse year-round daylight saving time" or "morning-daylight saving time."

Finally, in the spring of 1966, a bill standardizing DST was brought to a final vote in Congress. It called for a national DST period from the last Sunday in April to the last Sunday of October. The bill was approved by voice vote in the Senate on March 29, and the next day the House passed it by a large majority, 281 to 91. Opponents fought the bill to the bitter end, sending telegrams to President Johnson urging a veto. Marshall Fine of the National Association of Theatre Owners wired the president that federally imposed DST "may well be disastrous, particularly in the case of outdoor theatres. Since showing of motion pictures cannot begin until dark, outdoor operation in many areas would be virtually impossible, [and it] would certainly prevent much of the family attendance upon which we depend for existence." But such appeals had little impact against the general consensus in favor of the bill, and on April 13, 1966, President Johnson signed the Uniform Time Act of 1966 into law.

89TH CONGRESS, PUBLIC LAW 89–387
UNIFORM TIME ACT OF 1966

An Act to promote the observance of a uniform system of time throughout the United States.

Be it enacted by the Senate and House of Representatives of the United States of America in Congress assembled, That: This Act may be cited as the "Uniform Time Act of 1966."

Section 2. It is the policy of the United States to promote the adoption and observance of uniform time within the standard time zones . . . To this end the Interstate Commerce Commission is authorized and directed to foster and promote widespread and uniform adoption and observance of the same standard of time within and throughout each such standard time zone.

Section 3. (a) During the period commencing at 2 o'clock antemeridian on the last Sunday of April of each year and ending at 2 o'clock antemeridian on the last Sunday of October of each year, the standard time of each zone . . . shall be advanced one hour . . . except that any State may by law exempt itself from the provisions of this subsection . . . but only if such law provides that the entire State (including all political subdivisions thereof) shall observe . . . standard time . . . during such period.

(b) It is hereby declared that it is the express intent of Congress by this section to supersede any and all laws of the States or political subdivisions thereof insofar as they may now or hereafter provide for advances in time or changeover dates different from those specified in this section.

(c) For any violation of the provisions of this section the Interstate Commerce Commission or its duly authorized agent may apply to the district court of the United States . . .

Section 7. As used in this Act, the term "State" includes the District of Columbia, the Commonwealth of Puerto Rico, or any possession of the United States.

Approved April 13, 1966

Time uniformity for the United States.

The Uniform Time Act (UTA) standardized the national DST period at six months, from the last Sunday in April until the last Sunday in October, but included a provision that any state could exempt itself by passing a state law. Thus each state was left with two options: It could observe daylight saving time statewide during the official DST period, or it could operate under standard time all year. The UTA also bestowed greater enforcement powers on the Interstate Commerce Commission, a move the ICC had been urging since at least 1931. A year later, on April 1, 1967, a new cabinet-level Department of Transportation (DOT) was created whose responsibilities included taking over all the powers and duties of the ICC related to time, including those it was charged with under the Uniform Time Act. The first period of uniform daylight saving time under the UTA was scheduled to begin just three weeks after the new department was formed.

Some states accepted the new law casually and took no action, thus accepting statewide DST by default. Some states left the decision to their citizens by holding a referendum on DST; in Colorado, for instance, voters had approved DST by a wide margin in November 1966. But many other states witnessed continued passionate debate on the subject.

In Georgia, State Senator Bobby Rowan angrily exclaimed to his fellow lawmakers, "Not since biblical times has there been a man who could change sunrise and sunset, but the bureaucrats are attempting to do it." The Georgia legislature eventually voted to exempt the state from daylight saving time if at least two of its adjoining states did the same. None did. The new state of Hawaii, situated near the equator and enjoying virtually unvarying day lengths throughout the year, quickly exempted itself from the UTA. In referenda held in November 1968, South Dakotans voted in favor of daylight saving time, and

Michigan voters split almost exactly in half, rejecting daylight saving time by a mere 490 votes out of over 3 million votes cast.

In Arizona, a state where summer daytime temperatures often exceed 110°F and that had never observed DST except during wartime, the legislature failed to enact an exemption for 1967 and Arizonans bombarded their legislators with complaints. "We must wait until about 9 P.M. DST to start any nighttime activity such as drive-in movies, moonlight rides, convincing little children it's bedtime, etc.," the editors of the *Arizona Republic* groaned. "And it's still hot as blazes!" When an exemption passed the following year, the *Republic* observed that "drive-in theaters, the parents of small children, the bars, the farmers and those who do business with California" would be happy that Arizona was staying on standard time, whereas "power companies, the evening golfers, the late risers, and the people with business interests on the Eastern seaboard" would not. One legislator summed up the view of many residents: "The last thing Arizona needs in the summer is an extra hour of daylight."

ISLANDS OF TIME

By March 1971, four states had exempted themselves from advanced time: Arizona, Indiana, Hawaii, and Michigan. Indiana's situation was the most complex. The Uniform Time Act's provision that each state must make a single statewide decision on DST was contentious in states split by a time zone boundary—and it was most contentious in Indiana.

There was a long history of resistance to DST in Indiana. In March 1949, its House of representatives had considered a bill to ban the use of daylight saving time that resulted in the first filibuster in the one-hundred-year history of the Indiana legislature. Pro-DST urban forces tried to talk until time would run out on the session at midnight, but Herbert Copeland, a leader of the rural faction, leaned over the gallery railing and forced the official clock backward, which broke it and caused it to stop at 9 P.M. After hours of posturing on the floor of the House, at 3:30 A.M. the bill came to a vote and passed by a wide margin. The new law required Indiana's state and local governments to follow Central Standard Time but allowed businesses to adopt DST if they preferred.

Under the original time zones established in 1918, all of Indiana was in the Central time zone. But a series of westward zone-boundary shifts over the years had pushed some Indiana residents into the Eastern time zone. In April 1969, after eighteen months of complex proceedings, another shifting of time zone boundaries extended the Eastern time zone farther west to include almost all of Indiana, except for two small but well-populated pockets that remained on Central Time: the six northwestern counties near the city of Gary, which wanted to retain the same time as Chicago, and the six southwestern counties around Evansville, which wanted the same time as nearby Illinois and Kentucky.

But with the new time zone boundaries, the state's citizens were compelled by the Uniform Time Act, to "make a cruel choice," in the words of Indiana Congressman J. Edward Roush. Either they could "adopt daylight time, which gives eastern Indiana a kind of 'double daylight saving time,'" so that the western corners could enjoy Central DST to match their neighboring states, or they could force western Indiana to "accede to Central Standard year round so as not to saddle the east with daylight time." Recalling the infamous Steubenville-to-Moundsville bus ride, Rouch pointed out that a summer traveler going from eastern Kentucky to Illinois via Indiana would fall under four different times—Kentucky's Eastern Daylight Time, then eastern Indiana's Eastern Standard Time (an hour earlier), then southwestern Indiana's Central Standard

Time (another hour earlier), and finally Illinois's Central Daylight Time (an hour later).

In 1971, the Indiana legislature followed the desires of the
majority and passed a UTA exemption law. When DST began
nationwide in April, all of Indiana remained on eastern standard time, and the two western corners of the state on Central
Standard Time became small "Islands of Time," surrounded by
areas whose time was one hour later—Central Daylight Time to
the west and eastern standard time to the east.

Primarily in response to the situation in Indiana, in 1972 Congress passed an amendment to the Uniform Time Act allowing a
state that fell in more than one time zone to exempt from daylight
saving time either the entire state or just the part in one of its time
zones. The amendment was signed into law by President Richard
Nixon on March 30, 1972. Although eleven other states were
located in two time zones and Alaska had four time zones, only
Indiana utilized the partial exemption offered by the amendment.

92ND CONGRESS, PUBLIC LAW 92-267
UNIFORM TIME ACT AMENDMENT

An Act to amend the Uniform Time Act to allow an option in the adoption
of advanced time in certain cases.

*Be it enacted by the Senate and House of Representatives of the United
States of America in Congress assembled,*

That . . . the Uniform Time Act of 1966 . . . is amended by . . . inserting the
following: . . . any State with parts thereof in more than one time zone may
by law exempt either the entire State . . . or may exempt the entire area of
the State lying within any time zone.

Approved March 30, 1972

Amendment to the UTA for multiple-time-zone states.

As the United States adjusted to standardized national DST during the 1960s, worldwide use of daylight saving time was spotty and frequently in flux, as it had been since the postwar period. Some countries, such as Britain, had advanced time regularly each summer, while others used DST only when there was a specific need, such as during an energy shortage. Still others experimented with DST from year to year. Italy had remained on standard time for over twenty years due to the unhappy association of DST with wartime. But in May 1966, it adopted daylight saving time, or *ora legale* ("legal time"), as a way to reduce both traffic accidents and electricity bills.

In the Americas, in October 1963, President João Goulart of Brazil instituted *hora de verão* ("summer time") in the country's southern states for the first time in ten years. This was done in an effort to conserve energy since hydroelectric power, the region's major source of energy, decreased significantly during the hot, dry summer. When the need for conservation continued into December, President Goulart decreed *hora de verão* until March for all of Brazil. In June 1964, Cuba put its clocks ahead one hour and went on DST for the first time since World War II. And in 1966, the Canadian province of Saskatchewan essentially opted out of DST by establishing year-round Central Standard Time in its eastern districts, while allowing its western districts to choose between year-round Central Standard Time and Mountain Time with DST. After a few years, almost the entire province followed year-round Central Standard Time.

DAYLIGHT SAVING TIME SNAPSHOT: THE 1960S	
Africa	Portugal
Egypt	San Marino
Morocco	Turkey
Sierra Leone	United Kingdom
Syria	
	North America
Asia	Canada
Hong Kong	Cuba
Macao	Dominican Republic
Republic of China (Taiwan)	United States
South Korea	
	Oceania
Europe	Australia
Iceland	
Ireland	**South America**
Italy	Argentina
Malta	Brazil
Norway	Uruguay
Poland	

DST between the years 1960 and 1969.

By the late 1960s, Britain and Italy were the only countries in Western Europe that continued to use DST. When summer time was in effect, Britain's time agreed with most of Europe's (because most of the continent was in the next time zone, which was one hour later), but it fell one hour behind Europe in winter. Many British business leaders felt this situation put them at a competitive disadvantage, and some Britons felt it went against a spirit of pan-Europeanism at a time when Britain was considering entry into the European Common Market. Soon the idea of synchronizing Britain's time with the Continent's gathered momentum.

In 1968, despite significant opposition, the Labour-dominated

Parliament approved a three-year experiment, called British Standard Time (BST), which extended summer time for the entire year. With year-round DST, Britain's time was now the same time as that of most of Europe. Many in Scotland and Northern Ireland, where some midwinter sunrises were now being delayed until 9:45 A.M. or later, vehemently opposed BST, but a public opinion survey found it had the backing of a slight majority of the overall British population.

Opponents of BST soon called for its repeal, citing reports of children injured in automobile accidents on their way to school in the morning; studies by government analysts, however, showed a statistical decrease in the total number of accidents. Leading the opposition in Parliament was a longtime DST opponent, Sir Alan Patrick Herbert, now eighty years old. "The correct name for British Standard Time is Central European Time," Herbert argued, "and there is really no more to be said. (This is not insularity speaking, but astronomy and geography.) It is the time, appropriately, of eastern Poland and Yugoslavia, but is at present forced by the British Parliament upon the west of Ireland."

Finally, in December 1970, the House of Commons, in a "free vote" without any party control, voted overwhelmingly to discontinue British Standard Time. To reestablish DST for the following summer, Parliament passed the Summer Time Act of 1972. It contained a complex new definition of the summertime interval as "the period beginning at two o'clock Greenwich Mean Time, in the morning of the day after the third Saturday in March or, if that day is Easter Day, the day after the second Saturday in March, and ending two o'clock, Greenwich Mean Time, in the morning of the day after the fourth Saturday in October." *The Oxford Companion to the Year* points out the great forethought that was put into this definition, since the third Saturday in March will not be the day before Easter until the year 2285.

Meanwhile, back in the United States, as the 1960s ended DST played a surprising role in one of the most bitterly contested issues of the era. During the Vietnam War, young men were drafted for military service according to a lottery system whereby each year a government computer randomly assigned a national Random Sequence Number (RSN) from 1 to 366 to each of the 366 possible birth dates. All eligible individuals born on the date corresponding to RSN 1 were drafted first, then those born on the date corresponding to RSN 2 were drafted next, and so on until the required quota of draftees was filled. In Wilmington, Delaware, a young man named Chester Brinton had been born at 12:03 A.M. on August 12, 1948. In the 1970 draft, August 12 was assigned an RSN of 142 — a relatively low number — and Brinton was ordered by his local draft board to report on December 4, 1970, for induction.

But Brinton found a loophole. Under a 1923 Delaware law still in force when he was born in 1948, Eastern Standard Time had been declared "the standard time for all official and legal purposes in the state." Although he had been born during Eastern Daylight Time in Wilmington, the recording of births was a state function and Brinton argued that his official time of birth was not 12:03 A.M. Eastern Daylight Time on August 12, but rather 11:03 P.M. Eastern Standard Time on August 11. And August 11 had been assigned RSN 342, a very high number indeed.

When the draft board refused to recognize this revised birth date, Brinton took his claim to U.S. District Court. Judge Caleb Wright ruled that although it was surely not the intent of the Selective Service lottery procedures to allow an individual to choose the date upon which his RSN would be based, in Brinton's case the Delaware state standard time statute clearly superseded the local Wilmington daylight saving time ordinance. Judge Wright decided in Brinton's favor—and the 1970 draft never reached RSN 342, or Chester Brinton. The state of Delaware, however, quickly revised its laws to eliminate this loophole.

Brinton's case aside, in the early 1970s the six-month "last Sunday in April to the last Sunday in October" DST period continued to be popular in the United States. With the 1972 amendment addressing the plight of states in multiple time zones, it seemed that the daylight saving time controversy was finally resolved to the satisfaction of most. But soon a new nationwide crisis would once again make DST an issue of critical national debate.

Chapter Eight
Daylight for Oil

We burn daylight.

—Thomas Kyd, *The Spanish Tragedy*

In 1973, the outbreak of war would again have a major influence on the story of daylight saving time — but this time indirectly. On October 6, which marked Yom Kippur (the Day of Atonement), the holiest day of the Jewish calendar, Egyptian and Syrian forces launched a surprise attack against Israel on two fronts in an effort to force Israel to surrender previously captured territories. Thus began the fourth Arab-Israeli War, known as the Yom Kippur War or the October War. When Western allies, including the United States, aided Israel with supplies, the oil-producing Arab members of the Organization of Petroleum Exporting Countries (OPEC) retaliated. On October 17, Saudi Arabia, Libya, and other Arab states, which together controlled a large portion of the world's oil, cut oil production, raised prices, and declared a complete embargo on oil exports to the United States.

As a result, and without any advance warning, the United States encountered its first prolonged peacetime energy shortage. The government had the task of quickly establishing policies that would reduce energy consumption with minimal social and economic disruption. Only one day after the oil embargo was announced, advocates of extended daylight saving time reintroduced previously proposed bills to establish year-round DST, but now these bills included a statement of purpose related to the impending energy crisis. When the country switched from daylight saving time to standard time on October 28, the *New York Times* predicted that such a time change "may never happen again."

AN EMERGENCY ACT

Faced with this unprecedented crisis, President Richard Nixon addressed the nation on the prospect of "the most acute shortage of energy since World War II." After describing a number of executive measures he would take immediately to ease the situation, he promptly encouraged Congress to pass emergency legislation to "authorize an immediate return to daylight saving time on a year-round basis."

With strong presidential support and an anticipated shortage of three million barrels of oil per day, 17 percent of demand, Congress quickly went to work. Senator Warren Magnuson of Washington declared the continuation of standard time an "unnecessary energy consumption" and listed the advantages of year-round DST: "It is a simple procedure; it can be instituted quickly; it will save substantial amounts of energy; it will affect every person in the nation by involving them in an ongoing act of energy conservation; it will decrease crime; it will decrease traffic accidents; it will add to the personal convenience and safety of a large majority of citizens; and it has widespread public support." The Department of the Interior estimated energy savings from winter daylight saving time at "about one percent—a very significant amount." Opponents, as expected, came primarily from rural districts, and old animosities were revived. Representative H. R. Gross of Iowa, a longtime DST foe, called the proposal "the golfer's delight."

As Congress considered several conservation options, the energy situation continued to worsen. Heating oil prices rose to new levels and gasoline availability decreased dramatically, with prices jumping 40 percent and long lines at gas stations becoming routine.

While Congress continued to debate, the good citizens on Block Island, Rhode Island (population five hundred), took action on their own. On November 12, the town council of New Shoreham, the only municipality on the isolated island, voted to go on daylight saving time to conserve energy. This patriotic move put the residents of the tiny island at least one hour ahead of everyone else in America. Yet because of the island's remote offshore position and slow-paced lifestyle, the change had little effect on its residents' lives. When town official Herbert Whitman was asked, "What if the rest of the country refuses to go along with Block Island Time?" he replied, "Oh, phooey on them!"

About six weeks after the oil embargo was launched, the year-round DST bill passed the House 311 to 88 and a week later the Senate followed with a vote of 68 to 10. The large majorities included some reluctant supporters from Farm Belt states. The Emergency Daylight Saving Time Energy Conservation Act of 1973 provided for nationwide year-round daylight saving time for a trial period of about two years—from its effective date, January 6, 1974, until the usual end of the DST period in October 1975. On December 15, 1973, President Nixon signed the bill into law, declaring that whereas many actions taken to meet the energy crisis required inconvenience and sacrifice, "daylight saving time on a year-round basis . . .

will mean only a minimum of inconvenience and will involve equal participation by all. Unlike many of our other initiatives to deal with the energy crisis, . . . these savings will not require research, new technology, diplomacy, or exploration."

93RD CONGRESS, PUBLIC LAW 93–182
EMERGENCY DAYLIGHT SAVING TIME ENERGY CONSERVATION ACT OF 1973

An Act to provide for daylight saving time on a year-round basis for a two-year trial period . . .

Be it enacted by the Senate and House of Representatives of the United States of America in Congress assembled,

That this Act may cited as the "Emergency Daylight Saving Time Energy Conservation Act of 1973."

Section 2. The Congress hereby finds and declares
(1) that the United States faces severe energy shortages, especially in the winter of 1973–1974 and in the next several winters thereafter;
(2) that various studies of governmental and nongovernmental agencies indicate that year-round daylight saving time would produce an energy savings in electrical power consumption;
(3) that daylight saving time may yield energy savings in other areas besides electrical power consumption. . . .

Section 3. (a) Notwithstanding the provisions of . . . the Uniform Time Act of 1966, . . . the standard of time of each zone . . . shall be advanced one hour.

Approved December 15, 1973

DST for the energy crisis.

DARK WINTER MORNINGS

When the Emergency Daylight Saving Time Act took effect early on Sunday morning, January 6, 1974, the United States had its first taste of winter daylight saving time since World War II — to

predictably mixed reviews. The *Boston Globe* reported that the
city greeted the switch to winter DST with a widespread "ho-
hum, so what," but in many parts of the country, millions of
people grumbled as they woke up in what seemed to be the
middle of the night. In New York City, Sunday morning church
attendance was off; at the Chicago Zoological Park, zookeepers
had to wake up the wolves for their Sunday dinner of ground
turkey and horsemeat; and in Houston, 450 longshoremen
staged a five-hour boycott because a revised work schedule
under DST cost them an hour of overtime pay. But in Okla-
homa City, six-year-old Allen Cope was excited about the time
change: "I like it because it will give me more time to learn to
ride my bike."

As before, some of the strongest objections to DST came
from parents anxious about their children's safety as they left
for school in the dark. The extremely late sunrises under winter
DST made the problem that much worse. Farm families feared
accidents on the dark rural roads as their children waited for
the school bus, while in Philadelphia an angry mother rose at
a neighborhood school board meeting to shout: "It's dangerous
for our kids to go to school through pitch-black streets full of
abandoned buildings. Mothers, stop dragging your feet! Open
schools later in the day or our children are going to get hurt!"
Attempting to minimize the problem, some school districts
pushed their starting times back one hour, but many schools
were unable to do so because of union contracts, the after-
school commitments of school bus drivers, or the inconvenience that any change would cause working parents.

Many school authorities recommended that children wear
light-colored clothing and reflector patches to make themselves more visible to early-morning motorists. Officials in
Memphis, Tennessee, distributed more than seven thousand

yards of reflective tape to their twenty-one thousand students. The Connecticut General Life Insurance Company gave away "hot dots," reflective circles that adhered to children's clothing, and schools in Dallas, Texas, spent $9,000 on reflector signs for school crosswalks. All over the country, children illuminated their way to school with flashlights, lanterns, and even candles (reversing at least some of the candle-saving benefits of early awakening that Benjamin Franklin had envisioned).

"HONEST, MOM, WE *TRIED* TO GO TO SCHOOL, BUT WE COULDN'T FIND IT"

Problems of winter DST.

Despite such precautions, within two weeks of the start of winter DST, a sixteen-year-old boy on his way to school was run over by a car in suburban Chicago. Schoolchildren also died in predawn accidents in Virginia, Ohio, Louisiana, Pennsylvania, and California. Some police authorities and politicians blamed the darkness, at least in part, for these and a number of other traffic accidents involving children.

The fiercest controversy over motor-vehicle accidents involving children occurred in Florida. In the first weeks of winter DST, eight Florida children on their way to school were killed by cars, compared to only two in the same period of the previous year, causing a local TV commentator to declare that winter daylight saving time had proved to be "Daylight Disaster Time." Governor Reuben Askew urged the Florida legislature to reinstate standard time in the state declaring, "The inescapable conclusion is that darkness had a great deal to do with the predawn deaths." Although several legislators favored this move, others felt that a change of time in Florida alone would disrupt the state's industries, to which the governor retorted, "Any amount of disruption in commerce would be small indeed when compared to the life of even a single child." After heated debate, the legislature supported federal legislation, introduced by Florida's congressional delegation, to repeal winter DST nationally or at least allow an exemption for their state.

Around the country, however, most officials were reluctant to blame the accidents on daylight saving time, noting weather and other factors. Cities such as Buffalo and Ithaca, New York, and Independence, Missouri, reported fewer accidents in January 1974 than in the same period of the previous year, as year-to-year weather differences seemed to have more impact than DST. At congressional hearings, a Department of Transportation official pointed out that rather than causing additional motor-vehicle accidents, winter DST might well mean fewer total accidents because evening accidents would be reduced by the extra hour of evening daylight.

As for the legislation's role in reducing energy consumption, power companies around the country generally reported small decreases in energy use in the first weeks of winter DST. Los Angeles estimated a savings of about 1 percent, and the Bonneville

Power Administration, supplying much of the Northwest's power, reported a drop in energy use of "less than one percent." Dallas put the savings at "perhaps one-half of one percent," while Chicago reported a savings of less than 0.1 percent.

Within weeks of the new law's passage, public opinion of year-round DST had gone from 76 percent positive to 51 percent negative, and many bills were soon introduced in Congress to decrease or eliminate the extended DST period. "It's time to recognize that we may well have made a mistake," contended Iowa Senator Dick Clark in February 1974. But as the winter drew to an end, the urgency to reexamine year-round DST abated. Then, on March 18, 1974, the Arab countries ended the oil embargo against the United States, which eventually led to the end of the critical shortages of oil. But the issue of energy conservation remained a vital concern, as the price of oil had increased dramatically compared to what it had been before the embargo.

The following summer, Congress reviewed the previous winter of daylight saving time, and a Department of Transportation study analyzed the effects of the experiment. Acknowledging that DST effects could not be easily separated from other changes, such as voluntary reductions in the use of lighting, the DOT concluded that year-round DST probably resulted in a decrease in electricity consumption on the order of 0.75 percent for January and February and 1 percent for March and April. On the controversial subject of traffic accidents, the DOT found no significant overall change. Although schoolchildren may have been at greater risk in the morning, they were at less risk in the early evening. Nonetheless, Governor Askew noted that the number of morning traffic deaths for Florida schoolchildren had shot up in the January-April period compared to the previous year, while evening fatalities had gone down only slightly. "There is no way to know for certain how many of these accidents resulted from the morning darkness of daylight time or how many were caused by the unusually heavy fog," he observed. "Unfortunately, government can do little about the early morning fog . . . but it can control the clock."

Opinion surveys found that daylight saving time was generally popular with the public, except for the months of November through February. People seemed to prefer their daylight in the evening—as long as they and their families did not have to wake up or leave the house in the morning in total darkness. On the basis of these findings, the DOT recommended continuing extended DST but eliminating DST in the four darkest winter months, November through February. This plan would still provide two extra months of DST, March and April, and it was overwhelmingly approved by the House of Representatives in a 383 to 16 vote. The Senate seconded the change by voice vote, and President Gerald Ford signed the amendment into law on October 5, 1974. The United States would return to standard time on October 27, 1974, and resume DST on February 23, 1975—two months before the usual date.

Still, there was dissent. DST backers wanted to make the eight months of advanced time permanent, while some critics wanted to make the period of DST even shorter—for example, the three months from Memorial Day to Labor Day. To determine the best DST option, Congress asked the Department of Transportation to assess the results of the two years of extended DST. This investigation would become the largest study ever made of the effects of daylight saving time and would focus especially on DST in March and April, the two months that were added to the standard DST period in 1975.

THE IMPACTS OF DST

The effects of daylight saving time are complex and difficult to isolate. In many cases DST produces two small and opposite impacts—one in the morning that is due to less daylight and one in the evening that is due to more daylight—making the net daily impact even more difficult to ascertain. Furthermore, these changes take place in the presence of much larger effects generated by seasonal patterns, weather, and other changes such as energy availability and the state of the economy.

Activities that are clock-determined are shifted with respect to daylight, resulting in a complex set of effects. There is a shift of one hour with respect to daylight for most office workers, factory workers, and other "clock-oriented" people. Morning and evening rush hours are shifted by an hour with respect to daylight, and after-work options are expanded. For "daylight-oriented" people, including farmers, construction workers, and many retirees, for whom light is more important than clock time, DST causes no change in primary activities. But there is a secondary effect when these people interact with the clock-oriented world. Farmers, for instance, have long experienced inconvenience when selling milk from cows that follow the sun to markets that follow the clock. Moreover, many activities are impacted by both the clock and the sun. For example, daylight may encourage more shopping trips by removing the fear of street crime or other nighttime dangers, but many shoppers also have other commitments that impose clock-time constraints.

Using a variety of analytical techniques, the DOT study assessed the impacts of DST in March and April and concluded that "modest overall benefits" might be realized in three primary areas—energy conservation, traffic safety, and reduced crime—by use of an eight-month DST system (March through October) rather than the Uniform Time Act's six-month DST system (May though October).

Most subsequent studies of the effects of daylight saving time, performed in the United States, Britain, France, Israel, Mexico, New Zealand, and other countries, have produced somewhat similar results to the DOT study, especially with regard to the benefits of DST for energy conservation and traffic safety.

A primary impact of daylight saving time is the reduction of energy consumption, and this has been the major impetus for numerous countries to adopt DST. Because factories, businesses, and government offices, among others, often open at a time when the sun has already risen but do not close until after sunset, a clock advance of one hour allows them to save significant energy for lighting. The extra hour of evening daylight saves most households one hour of electricity for evening lighting, and also draws people outdoors, cutting additional indoor energy use. This savings may be wholly or partially offset by additional lighting needed in the morning, but many people sleep through the hour of sunrise, whereas almost everyone is awake during the hour of sunset. DST also often reduces the daily peak needed for electricity production (when the least efficient power sources are used) by spreading out usage to later in the evening. The DOT concluded that the total electricity savings associated with DST amounted to about 1 percent in spring and fall, corresponding to national savings of forty to fifty megawatt hours per day.

DST also might affect home heating, air conditioning, and other forms of energy consumption. For example, the extra hour of light in the evening could cause an increase in recreational and shopping travel by automobile (and therefore an increase in gasoline consumption) that might not be offset by a corresponding decrease in the morning. On the other hand, more outdoor activities might save energy by decreasing the use of TV sets and appliances. The DOT did not detect any significant DST impact on these areas.

One benefit of DST.

Another major impact of DST is the reduction of motor-vehicle accidents and fatalities. Driving after dark is much more dangerous than driving in daylight, and while there are other factors, this difference results primarily from decreased visibility. Since DST makes evenings lighter and mornings darker, the evening accident rate should decrease, while the morning rate should increase, for drivers and passengers as well as pedestrians. Since evenings see significantly more traffic than mornings — often twice as much — the overall daily accidents might be expected to decrease under DST. And better visibility is all the more important when another element is considered: early-evening drivers are more likely than morning drivers to be tired or inebriated. Certainly, traffic-pattern changes, weather, and other factors also may play a role in the incidence of accidents, but a shift to DST would be expected to reduce total accidents. In fact, the DOT study found a 0.7 percent decrease in fatal motor vehicle accidents for March and April under DST as compared with standard time. The decline was small but important, corresponding to approximately fifty lives saved and two thousand injuries avoided for the two-month period.

On the heated topic of safety for schoolchildren, dark DST mornings increase the risk of accidents for children on their way to school. However, the extra light from DST in the late afternoon decreases the risk of accidents for children in activities such as riding bicycles, engaging in unsupervised outdoor play, or traveling as passengers in cars. The DOT study found that under DST in March and April, the increase in morning accidents seemed to be more than offset by the decrease in evening accidents. Despite these findings, one political fact was crystal clear: The news stories of the tragic deaths of young victims in morning accidents carried far more emotional weight than statistics showing that fatalities were avoided in the evening.

Another area of DST impact is crime reduction. People generally feel safer in the daylight, and many types of crime are believed to be influenced by lighting conditions. For example, more light in the evening decreases the opportunity for street crime against people returning home from work. The DOT study found that violent crime in Washington, D.C., was reduced by 10 to 13 percent during periods of daylight saving time.

One somewhat unexpected area of daylight saving time impact is radio broadcasting. Because the sun's radiation on the atmosphere decreases the propagation of AM (amplitude modulation) radio signals during daylight hours, as darkness falls the ionosphere—the region of the atmosphere beginning about thirty miles above the earth—can reflect AM radio signals back to the earth for much greater distances than during the daytime. These reflected signals can cause interference with other broadcasts on the same frequency. Consequently, more AM stations can operate during the day without interfering with one another than at night.

This phenomenon allows the licensing of hundreds of "secondary" daytime-only stations, authorized by the government to operate from sunrise to sunset on the same frequency as a full-time station. These are generally small-town stations serving a rural area. To avoid causing interference, secondary stations must shift their operations with the shift in DST daylight, losing one hour's morning audience and gaining one hour's evening audience. Since listeners often tune in to local stations in the morning for weather, traffic, and school information, and then listen to a larger station in the evening for regional and national news, most secondary stations experience a net audience loss. The DOT study concluded that the small loss in advertising revenues as a result of DST was of less concern than the public's loss of early-morning information—especially in rural areas where a secondary station was often the only early-morning local broadcaster available to the community.

Daylight saving time benefits many enterprises related to outdoor pursuits, and it also impacts a number of other economic areas, such as manufacturing, domestic trade, construction, and public transportation. Groups surveyed in these areas mildly favored DST or felt it had no effect. A shift of clock time under DST lengthens the overlap of U.S. business hours with Europe and shortens the overlap with Japan. A DOT analysis showed no DST effect on communications with Japan, but an increase in communications with Europe.

Overall, the public's view of DST remained favorable. A March 1975 survey found 51 percent of the public in favor of extending DST to eight months, 28 percent against, and the remainder neutral. Most people in favor mentioned the benefit of more light in the evening, while the major complaint of those opposed was "problems with children going to school in the dark." Given the relative popularity of DST and the conclusions of its major study, the DOT recommended increasing the standard six months of DST to an eight-month DST system for at least two more years.

The DOT study also investigated the best days to begin and end DST each year. The DST period under the Uniform Time Act was approximately centered on the summer months of the year, July and August. But the DOT noted that the times of sunrise and sunset in March and April, which had standard time, were generally similar to those in September and October, which had DST. This situation seemed unbalanced, since DST should treat similar daylight patterns equally.

Therefore, the DOT recommended that a day in spring and a day in fall with similar sunrises and sunsets should be treated the same—either both with or both without DST. But determining which days were "similar" was not obvious, because day length, sunrise time, and sunset time, surprisingly, do not correspond to each other. In Houston, Texas, for instance, the earliest sunrise is on June 11, the longest day is June 21, and the latest sunset is on July 1. (This discrepancy is due to the use of mean time, which makes the clock times of sunrise and sunset lead or lag behind the actual sun.)

Considering all of these issues, the DOT proposed a "symmetrical" DST period based on sunrises, where days with the same sunrise time would either both have DST or both have standard time. Basing the DST period on sunrise time minimizes late sunrises—the major cause of DST's adverse effects. Various ways to choose the precise days to begin and end DST each year were considered. One idea was to ensure that DST was in force for specific events. DST on Halloween would provide more light for children on the streets, and DST on Election Day would encourage greater voter participation. In the end, however, since Americans had accepted the sunrises under the Uniform Time Act, the DOT proposed selecting a DST period in which the year's latest sunrise occurred at about the same time as under the UTA. For example, having DST from the third Sunday in March to the last Sunday in October would add more than a month of DST, and yet no area of the country would ever have a later sunrise than the latest it had under UTA.

THE RETURN OF THE UTA

The Emergency Daylight Saving Time Act expired on April 27, 1975, and the nation reverted to the Uniform Time Act's system of DST from May through October. Congress continued to consider several daylight saving time options, especially the DOT's proposal for eight months of DST. An attempt to establish a compromise seven-month DST period passed in the Senate, but in the House of Representatives, the bill was delayed by DST opponents whenever possible. The *Boston Globe* condemned this "daylight procrastination" and implored Congress to "show that it understands the difference between night and day." Further weakening the bill's chances was the fact that it reached the House at the end of the session, when a two-thirds majority was required to suspend the rules and pass a bill. A nearly empty House finally voted in favor of the bill, 11 to 10, but that slim majority was not enough, and the bill did not pass.

With the Uniform Time Act once again in effect, the country experienced the familiar six months of DST and six months of standard time. But, as always, the occasional anomaly arose. On April 24, 1977, the residents of De Land, Florida, dutifully set their clocks forward an hour for daylight saving time before retiring on Saturday evening. At exactly 2 A.M. that night, however, a Florida Power Commission substation lost power and stayed down for fifty-five minutes, cutting off electricity to a large section of the city. The next morning, De Land's electric clocks were about an hour slow, and its populace, unwittingly, was still on standard time. The only consolation residents had

DAYLIGHT SAVING TIME SNAPSHOT: THE 1980S	
Africa	Poland
Algeria	Portugal
Chad	Romania
Egypt	San Marino
Libya	Spain
Syria	Sweden
Tunisia	Switzerland
	Turkey
Asia	Union of Soviet Socialist Republics
China	United Kingdom
Hong Kong	West Germany
Iran	Yugoslavia
Iraq	
Israel	**North America**
Lebanon	Bahamas
Mongolia	Barbados
Republic of China (Taiwan)	Belize
	Bermuda
Europe	British Honduras
Albania	Canada
Andorra	Cuba
Austria	Dominican Republic
Belgium	Greenland
Bulgaria	Haiti
Cyprus	Jamaica
Czechoslovakia	Nicaragua
Denmark	United States
East Germany	
Finland	**Oceania**
France	Australia
Greece	New Zealand
Hungary	
Ireland	**South America**
Italy	Argentina
Liechtenstein	Brazil
Luxembourg	Chile
Malta	Paraguay
Monaco	Uruguay
Netherlands	
Norway	

DST during the years 1980 to 1989.

on arriving late by an hour for their Sunday morning activities was the knowledge that those who had forgotten to advance their clocks would arrive two hours late.

SOFTBALL AND BARBECUES

In each new Congress of the late 1970s and early 1980s, supporters introduced bills to lengthen the DST period. Such bills were passed by strong majorities in the Senate (in 1976) and in the House (in 1981), but none ever passed both in the same session of Congress. The benefits to the general public of additional DST did not seem persuasive enough to push a bill all the way through Congress, especially since some DST opponents held key congressional positions.

Then, in 1984, the business community got involved. Led by lobbyist James Benfield, a group calling itself the Daylight Saving Time Coalition was formed and backed by businesses and industry associations that stood to gain from additional DST. A leading member of the coalition was the Southland Corporation, parent company of the nation's 7-Eleven convenience stores. Southland felt that women were reluctant to approach their stores after dark, fearing undesirables lurking in the shadows. Fast-food companies, such as McDonald's, echoed this concern and also believed that more time for outdoor recreation would mean more softball players stopping in for a hamburger rather than cooking for themselves. A study by the Hardee's fast-food chain estimated that extending DST would increase sales by $880 per week per store.

Makers of barbecue grills and charcoal briquettes were also part of the coalition—more daylight would translate into more business. The same went for the Sporting Goods Manufacturers Association, which foresaw increased demand for softball, tennis, and golf equipment. Nursery owners wanted an extra hour of daylight in March and April to encourage gardeners to get an early start on spring planting. And more daylight would increase attendance at the establishments of the 1,700 members of the International Association of Amusement Parks & Attractions. As these and other businesses and industry groups joined the Daylight Saving Time Coalition, it grew to represent more than eight thousand companies with annual sales of over $100 billion. From a political standpoint, at least some of these companies were represented in virtually every congressional district. And many had an indirect effect— increased sales at McDonald's, for example, led to greater demand for Kansas beef and Idaho potatoes.

Candy manufacturers joined the DST Coalition with the very specific goal of extending daylight saving time to include Halloween. Many young trick-or-treaters gathering candy are not allowed out after dark, and an added hour of light could mean a big holiday treat for the candy industry. And at least one group in the coalition was not motivated by money: the Retinitis Pigmentosa Foundation Fighting Blindness, which represented the half million Americans who suffered from night blindness because of retinitis pigmentosa and other eye diseases. "The extra hour of daylight enables us to enjoy seeing for sixty minutes longer," explained Mindy Berman, speaking for the RP Foundation at a congressional hearing. "To someone slowly losing his or her vision, every second of sight is valuable."

In 1985, congressional advocates of daylight saving time, with the strong support of President Ronald Reagan and the Daylight Saving Time Coalition, introduced several bills to extend DST. The most popular of these was a proposal to establish DST from the first Sunday in April to the first Sunday in November, lengthening the DST period by three weeks in the spring and one week in the fall. Senator Slade

Gorton of Washington, chief sponsor of the bill, noted that the bill represented a rural-urban compromise: "We acknowledge the logic and benefits of a two-month extension, but we wish to cause a minimal amount of dislocation for those people, approximately 15 percent of our population, who live in the western regions of their time zones and who already have the benefits of later sunrises and sunsets."

Opposition in both houses of Congress came, as before, from farmers, from religious fundamentalists who continued to feel that DST was an offense against nature, and from representatives of areas at the western edge of a time zone. But DST advocates, led by Representatives Edward Markey of Massachusetts and Carlos Moorhead of California, supported the fight for the extension bill in the House, and, with the organized backing of the Daylight Saving Time Coalition, the bill passed in October 1985 by a vote of 240 to 157.

In the Senate, as with previous DST bills, resistance came from the Commerce Committee, and consideration of the bill dragged into 1986. In April, in an effort to bypass the committee, Senator Gorton added the DST extension as a rider to an unrelated, noncontroversial measure for U.S. fire prevention programs—a move reminiscent of the way DST repeal was handled after World War I. Gorton made an important concession by dropping the one-week extension into November. This maneuver greatly disappointed the candy manufacturers, who had wanted DST on Halloween, but the rest of the Daylight Saving Time Coalition was more concerned with the three-week spring extension. "The front end is the real money-maker anyway," Coalition Director Benfield pointed out.

The Senate passed the fire prevention bill with its DST provision by a voice vote, and the House passed it 386 to 28. On July 8, 1986, President Ronald Reagan signed the Federal Fire

---

99TH CONGRESS, PUBLIC LAW 99-359
FEDERAL FIRE PREVENTION AND CONTROL APPROPRIATIONS ACT OF 1986

To authorize appropriations for activities under the Federal Fire Prevention and Control Act of 1974

Be it enacted by the Senate and House of Representatives of the United States of America in Congress assembled,
That . . .

Section 2. (a) The Congress finds
(1) that various studies of governmental and nongovernmental agencies indicate that daylight saving time over an expanded period would produce a significant energy savings in electrical power consumption;

(2) that daylight saving time may yield energy savings in other areas besides electrical power consumption. . . .

(b) Section 3. (a) of the Uniform Time Act of 1966 . . . is amended by striking "last Sunday of April" and inserting in lieu thereof "first Sunday of April." . . .

Approved July 8, 1986

Extending DST to April.

Prevention and Control Act of 1986 into law, including the additional three spring weeks of DST. With the three-week extension, the nation seemed to have found a balance that satisfied both pro- and anti-DST forces. Pressure for any change— to make the DST period longer or shorter—virtually disappeared for a long while, and daylight saving time became a generally accepted, noncontroversial part of daily American life.

Chapter Nine

Modern Times

I don't mind going back to daylight saving time.
With inflation, the hour will be the only thing I've saved all year.
—Victor Borge

The entire United States currently observes daylight saving time, except for Arizona, Hawaii, and the islands of Puerto Rico, the U.S. Virgin Islands, American Samoa, and Guam— all of which have chosen to stay on standard time all year.

As established by the Uniform Time Act and extended by the act of 1986, the almost-seven-month daylight saving time period continued in effect since 1987. The issue of DST remained quiescent in the U.S. for almost twenty years. But in the spring of 2005, with oil prices soaring and U.S. energy consumption growing, Congressmen Fred Upton of Michigan and Edward Markey of Massachusetts proposed an amendment to a mammoth Energy Policy bill that provided for a two-month extension to the daylight saving time period. Under their amendment, DST would begin the first Sunday in March and end the last Sunday in November, essentially including the entire months of March and November. "Extending daylight saving time makes sense, especially with skyrocketing energy costs," said Rep. Upton. Echoing previous arguments, Rep. Markey added that in addition to saving energy, the extension would result in fewer traffic accidents, less crime, more recreation time, and increased economic activity. And moreover, "daylight saving just brings a smile to everybody's face."

The Upton-Markey amendment was passed by the House of Representatives by voice vote with general acceptance. However, when the proposal reached the Senate, some opposition surfaced, primarily from a new quarter, the U.S. airlines. The airlines had never before played a major role in the daylight saving time debate, but they were now concerned that the DST extension would put the U.S. significantly out of sync with the time in foreign countries. A two month extension would, for example, result in seven or eight weeks each year when the U.S. had DST while Europe did not. At many foreign airports, U.S. carriers have established fixed landing and takeoff time "slots" for which they cannot make short-term changes. Having to keep to these time slots during the U.S. DST extension would cause significant disruption to the airlines' schedules, and they anticipated a loss of many millions of dollars due to scheduling problems and lost connections for overseas flights. Beyond the airlines, there was concern expressed by some parent groups, including the National Parent Teachers Association about children going to school on dark mornings. Also, Orthodox Jews were disturbed about the extension through November, which would delay their sunrise prayers. Others were concerned about the effect on electronic devices with clocks. Those devices that automatically adjust for DST would have to be reprogrammed or manually changed.

From farmers, traditionally the most vociferous DST adversaries, opposition to the extension was muted. While some farmers criticized the proposal, modern farm equipment and techniques have made farmers much less dependent on the sun, and major farm groups took no stand on the amendment. "It is not a big issue," declared Tom Thieding of the Wisconsin Farm Bureau Federation. "Maybe just a little colder and a little darker in the morning."

Because of opposition from the airlines and others, the DST amendment was not included in the Senate version of the bill, and thus it went to Conference Committee to reconcile the House and Senate versions. In the Conference Committee, Senator Larry Craig of Idaho led the opposition to the DST extension and, after some negotiation, the Committee agreed to a compromise. The extension was cut back so that the new daylight saving time period would be from the second week in March through the first week in November—adding a total of one month of DST instead of two and eliminating the darkest mornings of the original proposal. To allow the airlines, electronic device manufacturers, and others to have time to make arrangements for the new DST period, the change was delayed almost two years, to March 2007. Also, to determine the actual energy savings resulting from the DST extension, the Department of Energy was mandated to report to Congress on the impact of the DST extension on energy conservation, after which Congress could reconsider the extension.

The Energy Policy Act of 2005, including the DST provision, was passed by the House on July 28 and by the Senate the following day, and President George W. Bush signed it into law on August 8. Under the Act, starting in 2007 daylight saving time in the United States begins the second Sunday in March at 2 A.M. and ends the first Sunday in November at 2 A.M. This DST period is more symmetrical than under previous plans, with sunrise times at the beginning and end of the DST period less disparate. Also, it always covers Halloween and thus provides additional light that evening, which should allow more safety crossing the street for many trick-or-treaters.

CONTENTMENT, CONTROVERSY, AND CONFUSION

While the daylight saving time period in the United States has changed over the years, in most parts of the country the observance of DST has been widely accepted. But even now, many years after the passage of the Uniform Time Act, there are still places where controversy and confusion continue to surround daylight saving time.

Long a hotbed of DST controversy, Indiana has finally settled a passionate debate that raged in the state for over thirty years. After the passing of the Uniform Time Act's "two time zone" amendment in 1972, the state's Central time zone area observed DST, but the remainder of the state, in the Eastern time zone, followed year-round standard time. An additional complication existed in the five southeastern counties near Cincinnati, Ohio, and Louisville, Kentucky, which unofficially observed DST to keep in sync with those cities. Through the years, this complex situation made DST a perennially contentious subject in the state, and Indiana politicians often treated the topic gingerly. In a 1996 primary debate, gubernatorial candidate Rex Early firmly declared, "Some of my friends are for putting all of Indiana on daylight saving time. Some are against it. And I always try to support my friends."

But in 2005, a new governor, Mitch Daniels, made the passage of statewide DST one of his top economic priorities, claiming that Indiana's unusual situation discouraged corporations from locating in the state and hurt Indiana companies by making it difficult and confusing for out-of-state firms to do business with them. The battle over the Daniels DST bill brewed for four months, until finally, after several heated sessions and close votes in the legislature, the proposal garnered just enough votes to pass—establishing statewide DST in Indiana beginning in 2006.

At 2 A.M. on a Sunday morning, a large number of bars are still open, and the effect of DST can be controversial. In many locations where liquor laws call for a 2 A.M. closing time, bars are actually required by law to close at 1:59 A.M. to prevent an extra hour of drinking from being added when the clocks are set from 2 A.M. to 1 A.M. in October. (The extra hour is often added anyway—unofficially, of course.) In Athens, Ohio, the site of Ohio University, the bars usually stay open until 2:30 A.M. But at the April transition to daylight saving time, patrons are robbed of a half hour of drinking when the clock springs forward at 2 A.M. In 1997, this indignity led over one thousand students and other late-night partiers to gather outside the bars along North Court Street in the center of the city. In what was later officially called a "time change disturbance," the protestors chanted "Freedom!" and threw liquor bottles, eggs, and road flares at police attempting to disperse the crowd. In the melee that erupted, forty-seven people, including thirty-four Ohio University students, were arrested.

Daylight saving time often causes confusion in the northeast corner of Arizona, where the region's huge Navajo Indian reservation has opted to observe DST despite the fact that the state of Arizona stays on standard time all year. The Navajo Nation extends into Utah and New Mexico, both of which go onto daylight saving time in summer, so the portion in Arizona also follows DST to keep the same time throughout the reservation. But the situation gets even more complex. The Hopi Indian reservation is completely within the Arizona part of the Navajo reservation, but the Hopi, who are completely independent of the Navajo, choose to follow the state of Arizona, not their

Navajo neighbors, and stay on standard time all year. Consequently, during the summer the Hopi reservation, on standard
time, is surrounded by the Navajo reservation, on DST, which
makes up a sizable part of Arizona, on standard time, whose
every neighboring state is on DST.

The DST transitions cause some confusion each year for railroad passengers. To keep to published timetables, trains cannot
leave a station before the scheduled time. So when the clocks
fall back one hour in October, all Amtrak trains in the United
States that are running on time stop at 2 A.M. and wait one hour
before resuming. Overnight passengers are often surprised to
find their train at a dead stop and their traveling time an hour
longer than expected. At the spring DST time change, trains
instantaneously become an hour behind schedule at 2 A.M., but
they just keep going and do their best to make up the time.

In every U.S. congressional session new bills are proposed
to further extend daylight saving time, usually to include
March or early November (and thus Halloween). None has
passed, but new issues related to daylight saving time continue
to emerge. In 2001, California experienced major power
shortages and extended DST was considered by Congressional
committees as a means of reducing power consumption.
Although the crisis subsequently abated, it is clear that Congress is prepared to consider extended DST as a tool if a critical energy shortage hits all or a section of the United States.

WORLDWIDE WILLETT

William Willett would be happy to know that daylight saving
time is now employed in about seventy countries around the
world, including almost every major industrialized nation. Sunrises, sunsets, and day lengths of countries near the equator do not
vary enough during the year to justify the use of DST, but even

in those countries it is sometimes adopted, especially for energy conservation.

In 1996, all of the member countries of the European Union adopted a common daylight saving time period, from the last Sunday in March to the last Sunday in October. The United Kingdom's summer time period varied from Europe's for many years, but since 1996 it has been the same as the EU's, and in 2002, Parliament permanently set Britain's summer time to match the continental plan. There have been several proposals for the United Kingdom to move to year-round summer time—essentially, to adopt Central European Time. Another proposed plan, called "Single/Double Summer Time," would have double summer time in the summer and regular summer time in the winter, as was used during World War II. Neither of these alternatives has generated enough support to pass in Parliament.

Most of Canada now employs the same DST period as the United States, but nearly all of Saskatchewan and a few other areas stay on standard time all year. In the spring of 1988, the Canadian province of Newfoundland (including Labrador) experimented with double daylight saving time (DDST) to capitalize on the long hours of summer sunlight available at the province's northern location. While DDST had been utilized before in Newfoundland (as A. P. Herbert had discovered), Britain, and elsewhere, this may have been the first instance when clocks were set ahead two hours at one time; previously DDST had added one additional hour on top of year-round DST. Many Newfoundlanders welcomed DDST, delighted with the extra-long period of daylight available after work for recreation, sports, gardening, or just relaxing. And the local musical group Simani celebrated it in song:

DOUBLE DAYLIGHT SAVING TIME

Uncle Tom's watch used to shine in the dark
But he gave her a weegee [toss] out over the wharf.
Don't need 'er now boys, Uncle Tom said,
Since the new daylight savin's time scheme was brought in.

Who thought it up boys, God love his heart.
'Tis wonderful, seein' our summer's so short.
With twice as much sunshine, sure Lord bless us all
I allow we'll be burned up as black as the coal.

Double your pleasure, double your fun
For half of the year we'll have double the sun.

But DDST also brought problems. It made Newfoundland's time as different from Ottawa's as it was from London's, and there was an hour less time for businesses to communicate with the rest of Canada. Of equal or more importance to many Newfoundlanders was the disruption to network TV shows, and televised West Coast sporting events often started well after midnight, leading many a bleary-eyed Newfoundlander to gripe about falling asleep during an important Stanley Cup hockey game. The situation grew worse after summer when schools opened; parents found it difficult to get their children to sleep on school nights and many children traveled to school in the dark. One October morning, more than 1,400 students in Labrador carried flashlights to school to dramatize the danger. After the DDST period ended, a provincial survey showed that while a majority favored some form of DDST, there was no consensus as to the preferred period. John Butt, the provincial minister of culture, concluded that "to implement DDST once more with no real consensus would produce chaotic results, causing inconvenience and confusion." Consequently, he declared that double daylight saving time would not be observed the following summer—and it never was again.

DST has now spread to every continent. In the Southern Hemisphere, in countries such as Chile, Namibia, and New Zealand, DST is observed from September or October to March or April. Even in Antarctica, where there is no daylight in the winter and months of twenty-four-hour daylight in the summer, many research stations observe daylight saving time anyway—to keep the same time as their supply stations in Chile or New Zealand.

Controversy over daylight saving time continues to erupt from time to time in countries around the world. In 1993, Brazil experienced a headline-making political battle over daylight saving time when its *hora de verão* (DST) period ended in February and Cesar Maia, the mayor of Rio de Janeiro and an economist, decided his city should remain on daylight saving time to save energy and stimulate after-work shopping. The federal government refused to go along with Maia's plan, and Rio's businesses fell into turmoil for two days until Maia backed down and had Rio's clocks reset. A similar clash broke out in Mexico in 2001 between two political rivals, President Vicente Fox and Mexico City mayor Andres Manuel Lopez Obrador, when the mayor decreed that his city would not accept the national government's DST, known as *horario de verano*. Obrador maintained, "You ought to get up when the rooster sings and go to bed when the cricket sings." Federal officials argued that DST saved hundred of millions of dollars in energy costs and helped Mexico's businesses coordinate their activities with those of their main trading partners in the United States. When Fox offered to shorten the daylight saving

DAYLIGHT SAVING TIME SNAPSHOT: TWENTY-FIRST CENTURY

Africa
Egypt
Namibia

Asia
Armenia
Azerbaijan
Georgia
Iran
Iraq
Israel
Jordan
Kazakhstan
Kyrgyzstan
Lebanon
Mongolia
Pakistan
Palestinian Territories
Syria

Europe
Albania
Andorra
Austria
Belarus
Belgium
Bosnia and Herzegovina
Bulgaria
Croatia
Cyprus
Czech Republic
Denmark
Estonia
Finland
France
Germany
Greece
Hungary
Ireland
Italy
Latvia
Liechtenstein
Lithuania
Luxembourg

Macedonia
Malta
Moldova
Monaco
Netherlands
Norway
Poland
Portugal
Romania
Russia
San Marino
Serbia and Montenegro
Slovakia
Slovenia
Spain
Sweden
Switzerland
Turkey
Ukraine
United Kingdom

North America
Bahamas
Bermuda
Canada
Cuba
Greenland
Mexico
United States

Oceania
Australia
Fiji
New Zealand
Tonga

South America
Brazil
Chile
Paraguay

Antarctica
Antarctica

DST since the year 2000.

time period by two months but refused to cancel it, he touched off a constitutional crisis. Eventually the Mexican Supreme Court stepped in and ordered the city to follow Mexico's national DST policy.

DST has been a continuing issue in the state of Queensland, in northeastern Australia, which has maintained standard time while the rest of Australia's east coast went onto DST. Businesspeople in Queensland, particularly those on the Gold Coast tourist strip, argue that DST is essential if they are to stay in sync with business partners, customers, and tourists from Sydney and Melbourne. In 2001, the Gold Coast even tried to create its own "time zone" by opening businesses an hour early, but government services and schools kept to standard time and the plan was abandoned after a year of confusion.

In Israel, DST is a long-simmering issue that has caused continuing contention between religious and nonreligious groups. Religious Jews, who must wait for sunrise to say early-morning prayers, want earlier sunrises, whereas most secular Israelis generally prefer the extra hour of evening daylight. In 1986 the dispute resulted in a demonstration outside the home of Israel's interior minister, Yitzhak Peretz. Protesters filled his street, where they unleashed a barrage of cuckoo clocks, alarm clocks, and bells. They were soon joined by hordes of counter-demonstrators. In Israel such political battles were usually won by religious groups, and for many years Israel had a shorter period of DST than most other countries.

On September 5, 1999, a stolen Fiat Uno wound its way from Dabburiya, Israel, toward the big city of Haifa, while at the same moment a stolen Audi was on the road to Tiberias. In each car were two young men, Israeli Arabs from the village of Mash'had. The four had planned a coordinated terror attack. They got the idea for the attack when they noticed that Israeli officials failed to examine the bags put in the luggage compartments of intercity buses. Their plan was to go out in two-man teams, one team to Haifa and one to Tiberias. A member of each team would board a bus to Jerusalem and place baggage containing explosives in the luggage compartment. His accomplice would follow the bus in a car. At some stop en route, the bomber would disembark and jump into the car with the accomplice, and the two would flee before the bomb exploded. Both bombs, each with fifteen kilograms (about thirty pounds) of explosives, had been set to explode on Sunday, September 5, at 6:30 P.M. — simultaneously, for maximum effect.

At about 5:30 P.M. on the appointed day, two of the young men, Ibrahim Salah and Krayem Nazal, drove in the Fiat to the Central Bus Station in Haifa. Salah stopped to buy a drink while Nazal parked the car. They could easily make the next bus to Jerusalem, plant the bomb, and have the bomber get off the bus well before 6:30. But all of a sudden there was a huge blast and intense smoke as the Fiat, with Nazal in it, exploded in the municipal parking lot on Yateh Street. The streets were instantly littered with broken glass and twisted metal, and four cars parked nearby caught fire. At almost exactly the same time, sixty miles away, as the second two-man team, Amir Masalha and Jad Azayza, drove at high speed down Elhadef Street in crowded downtown Tiberias, their Audi blew up in a massive explosion, shattering windows in nearby shops and apartments.

The police and security forces investigating the incidents pieced together what had happened. For many years, Israel had been resetting its clocks back from daylight saving time earlier than the rest of the world, and it had just fallen back to standard time the previous Friday, September 3. The Palestinian territories, however, following the neighboring Arab countries and most of the world, were still on daylight saving time. As the bombs were being assembled on the Palestinian West Bank, they had been set by the bomb makers to 6:30 P.M. according to Palestinian daylight saving time. But in Israel it would only be 5:30. Then when the bombs were smuggled into Israel, the four conspirators who received them, not realizing the importance of the hour difference, did not ask which time had been used to set the timing mechanisms. As a result, the bombs exploded one hour early—and instead of two busloads of Israelis, it was the three terrorists in the cars who were killed.

WILLIAM AND BENJAMIN

New arguments and concerns about DST continue to arise. As the number of household devices with clocks continues to increase, some have complained about the increasingly arduous task of resetting the clocks twice a year. The one-hour loss of sleep on the spring transition to DST disrupts the body's internal cycle to some degree. Many people are little affected by the change and consider it at most a minor annoyance, but others get a significant "jet-lag" effect, which sometimes lasts for several days, temporarily upsetting their daily routine. Some have claimed that this causes increased accidents for a few days after the transition, and there is even a theory that the hour lost to DST results in a one-day drop in stock market prices. On the other hand, many people relish the extra hour of sleep they get each autumn when the clocks are set back. Also, a longer DST period is thought to benefit sufferers of seasonal affective disorder, a form of depression related to the decrease of daylight in autumn and winter. And fire safety officials encourage the public to use the DST transition days as a reminder to replace the batteries in smoke detectors.

It is fitting that nowhere today is daylight saving time commemorated more elaborately than in Willett's own nation.

"You will fall backward. But eventually you will spring ahead."

The future of daylight saving time.

Every autumn, on the night when the clocks in England are turned back to Greenwich Mean Time, the students of Merton College at Oxford University perform the "Merton Time Ceremony." The ceremony, they say, is an attempt to preserve the regular passage of time, as the shift back from DST is made, by creating a small area where time stands still for one hour.

Wearing full academic dress and led by the "Keeper of the Watch," participants gather in the college's garden at 1:30 A.M., where they make two traditional toasts, one at 1:50 A.M.: "To good old time!" and another at 1:55, "Long live the counterrevolution!" Then the students walk to Fellows' Quadrangle, where at precisely 2:00 A.M., according to the Keeper of the Watch, they begin solemnly walking backwards, counterclockwise around the Quadrangle. As they walk in the dark, glasses of port in hand, many participants pause for a ceremonial twirl as they reach each corner of the quad. Every student makes at least two complete circuits of the quad, and at the end of the hour each makes a final circuit, as time—which they maintain had stopped for one hour—is once again restarted. And each year, when the universe has not slipped into limbo, they claim another success.

Now, one hundred years after William Willett first conceived his grand idea of shifting the clocks to save morning sunlight, his plan has become an international institution. For many, daylight saving time is the true demarcation of the seasons: The day the clocks are set forward each year, and not the vernal equinox, has become the much anticipated "first day of spring." Willett, who had such a frustrating time trying to convince his own countrymen of the merits of his idea, would certainly feel that his unrelenting efforts were fully justified, now that his scheme benefits well over a billion people on every continent of the globe.

And if Benjamin Franklin could have envisioned the effects produced by his idea of shifting human activity to make the best use of daylight—the problems, the benefits, the intriguing curiosities, the vociferous contentions, its impact on death row prisoners and train travelers, farmers and presidents, scientists and songwriters, and all of the other pieces that played a part in the story of daylight saving time—he surely would have been even more astonished than he was that morning in 1784 when he was awakened early and saw the sunlight streaming through his window.

SELECTED BIBLIOGRAPHY

The primary sources for this book were the many hundreds of newspaper articles chronicling the story of daylight saving time. This rich treasure trove of DST history was supplemented by articles in periodicals, government publications, books, and unpublished documents from archives and libraries.

NEWSPAPERS

Albany Argus (1883)
Arizona Republic (1968, 1969, 1974)
Athens (Ohio) *News* (1998)
Boston Daily Advertiser (1883, 1916)
Boston Evening Transcript (1937)
Boston Globe (1969, 1973–1976, 1985, 1986, 1993, 1997, 2000)
Boston Herald (1920)
Boston Herald American (1973, 1974)
Boston Post (1883)
Bromley (UK) Chronicle (1915)
Bromley (UK) Times (1915)
Buffalo Evening News (1946)
Chicago Sun-Times (1977, 1983)
Chicago Tribune (1918)
Christian Science Monitor (1987)
Cincinnati Enquirer (1909, 1916)
Cincinnati Times-Star (1909, 1914, 1920)
Clare (Ireland) *Champion* (2000)
Cleveland Plain Dealer (1914)
Clinton Iowa Herald (1945)
Des Moines Register (1943, 1945, 1962, 1964–1966)
Des Moines Tribune (1918, 1943, 1964, 1965, 1969)
Detroit Free Press (1915, 1949)
Fort William Times-Journal (1905)
Frankfurter Zeitung (Germany) (1916)
Grand Falls (Newfoundland) *Coaster* (1988)
Haaretz (Israel) (1999, 2000)

Halifax Chronicle-Herald (2003)
Halifax Herald (1916)
Hartford Courant (1923)
Harvard University Gazette (1999)
Houston Post (1920)
Indianapolis Daily Sentinel (1883)
Indianapolis Star (2004–2005)
Irish Examiner (1999)
Jersey Journal (1920)
Jerusalem Post (1986, 1999)
Los Angeles Examiner (1930)
Los Angeles Times (1918, 1930, 2001)
Madison (WI) *Capital Times* (2005)
Manchester (NH) *Union* (1916, 1920)
Manchester (NH) *Union-Leader* (1999)
Minneapolis Star (1965)
New York American (1918)
New York Commercial Advertiser (1883)
New York Evening Post (1919)
New York Herald (1883)
New York Sun (1915)
New York Times (1909, 1914, 1916–1927, 1929, 1932, 1933,
     1937–1939, 1946–1950, 1952–1959, 1962–1969, 1971–1975,
     1977, 1979, 1980, 1982, 1983, 1985, 1986, 1992, 2001, 2005)
New York Tribune (1883, 1909, 1916, 1919, 1920, 1922)
New York World (1919)
Newark News (1919)
Ohio University Post (2004)
Palm Springs Desert Sun (1946, 1947, 1950)
Pittsburgh Post-Gazette (1942, 1957, 1997)
Pittsburgh Sun (1919, 1922, 1927)
Port Arthur (Ontario) *News-Chronicle* (1942, 1951)
Raleigh News and Observer (1932)
Sacramento Bee (1999, 2001)
San Diego Union-Tribune (2001)
Seattle Times (1999)
Springfield (MA) *Republican* (1918)
St. John's Telegram (1987–1989)
St. Paul Pioneer-Press (1965)

The Times (London) (1908, 1909, 1911, 1914–1916, 1918, 1925, 1950, 1970, 2005)
Toledo Blade (1974)
Toronto Globe and Mail (1948, 1988)
Toronto Star (1988, 1998)
USA Today (2001, 2005)
Wall Street Daily News (1904)
Wall Street Journal (1990, 1996, 2005)
Washington Post (1922, 1963, 1998, 2001, 2002)
Webster City (Iowa) *Freeman Journal* (1941)

PERIODICALS AND JOURNALS

Accident Analysis and Prevention (1978)
The American City (1916, 1918, 1919)
American Journal of Public Health (1995)
The American Review of Reviews (1916)
Arizona Highways (1998)
Business Mexico (2002)
Bygone Kent (1994)
Cincinnati Magazine (1909)
Congressional Quarterly (1976, 1981, 1983, 1986)
Current Opinion (1918)
Fortune (1984)
Harpers Weekly (1883)
Israel Business Today (1993)
Journal of Commerce and Commercial (1991)
Journal of the American Institute of Electrical Engineers (1919)
Life Magazine (1947)
The Literary Digest (1916, 1918–1920)
Living Age (1909)
National Geographic (1990)
Nature (1906, 1908, 1909, 1911, 1916, 1917)
New England Journal of Medicine (1996)
Newsweek (1957, 1961–1965)
Official Airline Guide (1998)
Official Guide of the Railways (1964)
The Outlook (1908)
The Passing Show (1916)

People Weekly (1986)
Pictorial Weekly (1934)
Public Affairs (University of South Dakota) (1966)
Punch (1909, 1916)
Reader's Digest (1986)
The Reporter (1962)
Saturday Evening Post (1916)
Science (1909)
Scientific American (1884, 1908, 1909, 1911, 1916–1919, 1979, 1982)
Status Report of the Insurance Institute for Highway Safety (1993)
Survey (1918)
Technology Review (1977)
Time (1941, 1942, 1962, 1963, 1967, 1974)
U.S. News & World Report (1965–1967, 1974, 1996)
The Westminster Review (1909)
Yankee (1996)

GOVERNMENT PUBLICATIONS

United States
Congressional Record (1909, 1915, 1918, 1919, 1942, 1945, 1966, 1973, 1985, 1993)
House of Representatives Hearings (1920, 1941, 1971, 1974, 1976, 1981, 2001)
Senate Hearings (1916, 1917, 1985, 2001)
House of Representatives Reports (1919, 1941, 1942, 1945, 1949, 1966, 1971, 1973, 1974, 1985, 1986)
Senate Reports (1896, 1917, 1918, 1942, 1945, 1949, 1965, 1966, 1974, 1986)
U.S. Department of Transportation, *Standard Time in the United States* (1970)
U.S. Department of Transportation, *The Daylight Saving Time Study* (1974, 1975)
Congressional Research Service, *Daylight Saving Time* (1974, 1987, 1998)
California Energy Commission, *Effects of Daylight Saving Time on California Energy Use* (2001)
Public Papers of the President (1946, 1973)

U.S. Interstate Commerce Commission, *Annual Report* (1961)
U.S. Supreme Court Digest (1926)

United Kingdom
Report and Special Report from the Select Committee on the Daylight Saving Bill, H. M. Stationery office, London (1909)
Summer Time Committee, *Report of the Committee*, 1916, H. M. Stationery office, London (1917)
Board of Education, *Memorandum on the Effect of the Summer Time Act on the Health of School Children*, London (1922)
Secretary of State for the Home Department, *Review of British Standard Time*, Her Majesty's Stationery Office, London (1970)

Canada
A Green Paper on Changing Newfoundland and Labrador's Time Practices (1987)

Australia
Australian Government, Information Service, *Daylight Saving* (1995)

BOOKS

Aveni, Anthony, *Empires of Time*, Basic Books, New York, 1989.
Barnett, Jo Ellen, *Time's Pendulum*, Plenum Trade, New York, 1998.
Bartky, Ian R., *Selling the True Time: Nineteenth Century Timekeeping in America*, Stanford University Press, Stanford, CA, 2000.
Blackburn, Bonnie and Leofranc Holford-Stevens, *The Oxford Companion to the Year*, Oxford University Press, Oxford, 1999.
Blaise, Clark, *Time Lord: Sir Sandford Fleming and the Creation of Standard Time*, Pantheon Books, New York, 2000.
Boston Chamber of Commerce, *General Information on the Daylight Saving Plan*, 1916.
Boston Chamber of Commerce, *An Hour of Light for an Hour of Night*, 1917.
Bruce, William Cabell, *Benjamin Franklin Self-Revealed—A Biographical and Critical Study based Mainly on His Own Writings*, G. P. Putnam's Sons, New York and London, The Knickerbocker Press, 1923.

Churchill, Randolph, *Winston S. Churchill, Vol. II, Young Statesman 1901–1914*, Houghton Mifflin Co., Boston 1967.

Corliss, Carlton J., *The Day of Two Noons*, Association of American Railroads, Washington, D.C., 1952.

Cowan, Harrison J., *Time and Its Measurement*, The World Publishing Co., Cleveland, OH, 1958.

Curran, George Leo, Ps. D. & Taylor, Irene Hume, Ph. B., *World Daylight Saving Time*, Curran Publishing Company, Chicago, IL, Second Edition, 1935.

Davies, Robertson, *The Diary of Samuel Marchbanks*, 1947.

de Carle, Donald, *British Time*, Crosby Lockwood & Son Ltd., London, 1947.

Doane, Doris Chase, *Time Changes in Canada and Mexico*, AFA, Tempe, AZ, 1993.

Doane, Doris Chase, *Time Changes in the U.S.A.*, AFA, Tempe, AZ, revised 1997.

Doane, Doris Chase, *Time Changes in the World, except Canada, Mexico, USA*, AFA, Tempe, AZ, 1994.

Dowd, Charles F., *System of National Time and Its Application, by Means of Hour and Minute Indexes, to the National Railway Time-Table*, Weed, Parsons and Company, Albany, NY, 1870.

Dowd, Charles N., *Charles F. Dowd and Standard Time*, The Knickerbocker Press, New York, 1930.

Ferrell, Robert, ed., *Off the Record: The Private Papers of Harry S. Truman*, Harper & Row, New York, 1980.

Franklin, Benjamin, *Autobiography 1706–1757*.

Franklin, Benjamin, *Ben Franklin's Writings*, Library of America, 1987.

Franklin, Benjamin, *Poor Richard's Almanack*, 1735.

Garland, Robert, *Ten Years of Daylight Saving: from the Pittsburgh Standpoint*, Carnegie Library of Pittsburgh, 1927.

Goodman, Nathan, Ed., *Benjamin Franklin Reader*, Crowell Company, New York, 1945.

Goodsmit, Samuel A., and Claiborne, Robert, *Time*, Time-Life Books, New York, 1965.

Herbert, Sir Alan Patrick, *In the Dark: The Summer Time Story and the Painless Plan*, The Bodley Head, London, 1970.

Hillman, Mayer, *Time for Change: Setting Clocks Forward by One Hour throughout the Year(A New Review of Evidence*, Policy Study Institute, London, 1993.

Howse, Derek, *Greenwich Time and Longitude*, Philip Wilson Publishers, 1997.

Hughes, Robert W., *The Story of Daylight Saving Time*, Pittsburgh, 1936.

Kunz, George Frederick, *Saving Daylight for Economic and Preparedness Reasons*, Joint Conference Committee of the Engineering Societies, 1916.

Landes, David S., *Revolution in Time*, Belknap Press, Cambridge, MA, 2000.

More Daylight Club, *The Waste Of Daylight*, Detroit, MI, 1914.

O'Malley, Michael, *Keeping Watch: A History of American Time*, Viking Penguin, New York, 1990.

Pembor, Mary, *A Guide to Daylight Saving Time in the United States and Canada*, Chicago, IL, 1955.

Richards, E. G., *Mapping Time: The Calendar and Its History*, Oxford University Press, 1998.

Royal Greenwich Observatory, *Information Leaflets*.

Shanks, Thomas G., *The American Atlas*, ACS Publications, 1996.

Shanks, Thomas G., *The International Atlas*, ACS Publications, San Diego, CA, 1999.

Stephens, Carlene E., *Inventing Standard Time*, National Museum of American History, Smithsonian Institution, Washington, D.C., 1983.

Stevenson, Robert Louis, *A Child's Garden of Verses*, 1885.

Transportation Association of America, *Adjusting the Nation's Timepiece: The Story of the Uniform Time Act of 1966*, 1966.

Waymark, Peter, *A History of Petts Wood*, The Petts Wood and District Residents' Association, Bromley, U.K., 1990.

Whitrow, G. J., *Time in History*, Oxford Press, Oxford, 1988.

Willett, William, *The Waste of Daylight*, London, 1907, 1908, 1914.

SOURCES OF MISCELLANEOUS/UNPUBLISHED DOCUMENTS AND COMMUNICATIONS

Amtrak

Association pour L'Histoire des Chemins de Fer en France (Paris, France)

Boston Public Library

Bromley Central Library (Bromley, Kent, England)

The Builder Group Library (London, UK)
Chicago Motor Club
Cincinnati Historical Society
City of Rio de Janeiro, Brazil
Columbia University Library
Connecticut State Library
Detroit Municipal Reference Library
Detroit Public Library, Burton Historical Collection
Gerald R. Ford Library (Ann Arbor, MI)
Harry S. Truman Library (Independence, MO)
Harvard University Library
House of Commons Information Office (London, UK)
Library of Congress
Los Angeles Public Library
Lyndon B. Johnson Library (Austin, TX)
Manchester (NH) Historic Association
Middlesex Law Library (Cambridge, MA)
National Archives
National Cigar Museum
New York Public Library
Newfoundland and Labrador Legislative Library
Newfoundland Historical Society
Nixon Presidential Materials at the National Archives (College
    Park, MD)
Office of the Manhattan Borough Historian
Palm Springs Historical Society
Pathé Exchange Collection (Los Angeles, CA)
Public Library of Cincinnati & Hamilton County
Public Library of Steubenville (OH) and Jefferson County
Richard M. Nixon Library (Yorba Linda, CA)
Ronald Reagan Library (Simi Valley, CA)
Saskatchewan Library
State Historical Society of Iowa
Thunder Bay (Ontario) City Archives
Thunder Bay (Ontario) Historical Museum Society

ACKNOWLEDGMENTS

I would like to express my deep appreciation to the many people who have helped make this book a reality.

First, I am greatly indebted to my literary agent, Deirdre Mullane at the Spieler Agency, who provided excellent representation and did an outstanding job in helping develop and shape this book.

Much thanks, too, to my editor at Thunder's Mouth Press, Jofie Ferrari-Adler, for his enthusiastic support of this project and his first-rate editorial guidance.

During the extensive research, writing, and publication processes that went into this book, numerous people made significant contributions. I would like to thank them all, most especially the following: Bunny Ames; Linda Bailey of the Cincinnati Historical Society Library; James Benfield of the Daylight Saving Coalition; Bud Davidge of Simani; Paul Giannamore of the *Steubenville Herald-Star*; Dr. Tony Hyman; Elaine Levison; Congressman Edward Markey and his Legislative Director, Jeff Duncan; Melanie Martin of the Newfoundland Historical Society; Erin McCormick; Carolyn Morgan of the Newfoundland and Labrador Legislative Library; Dr. Norman Garber; Anthony Ody; Joanne Petrie of the U.S. Department of Transportation; Diane Raphael of the Office of the Manhattan Borough Historian; Jeffrey Robbins of Joseph Henry Press; Robert Selle; Dr. João Scalon; Dr. Ann C. Smith; Ben Steinberg of the Brookline Public Library; and David Stern. My great thanks to all of you and to the many others who contributed.

And finally, I couldn't have written this book without the help and encouragement, not to mention the detailed editorial recommendations, of my wonderful wife, Gail, and son, Michael. For their support and patience, I dedicate this book to them.

Credits

The author gratefully acknowledges everyone who gave permission for illustrations and song lyrics to appear in this book. Although every effort has been made to trace and contact copyright holders, if an error or omission is brought to our attention we will be pleased to correct the situation in future editions of this book. For further information, please contact the publisher.

Page 2. William Willett portrait from *The Waste of Daylight*, Nineteenth Edition, 1914.

Page 40. Charles Dowd's original plan for time zones from his *System of National Time*, 1870.

Page 41. Charles Dowd's time zone table from his *System of National Time*, 1870.

Page 46. Newspaper notice from the *Boston Advertiser*.

Page 58. Cartoon by George Whitelaw from *The Passing Show*.

Page 61. Advertisement from *The Times* (London).

Page 64. Advice on putting back the clock from *The Times* (London).

Page 67. Cartoon by Arthur Moreland from *The Passing Show*.

Page 86. Postcards from the National Archives.

Page 96. Newspaper notice from the *Halifax Herald*.

Page 99. Advertisement from the *Chicago Times*.

Page 104. Cartoon by Cy Hungerford from the *Pittsburgh Sun*.

Page 112. Cartoon by Lute Pease from the *Newark News*.

Page 116. Advertisement from the *Washington Post*.

Page 120. Cartoon by Cy Hungerford from the *Pittsburgh Sun*.

Page 137. Photographs of Willett memorial courtesy of Erin McCormick.

Page 138. Photograph of Daylight Inn sign courtesy of Erin McCormick.

Page 150. Cartoon © 1942, reproduced by permission of the *Pittsburgh Post-Gazette*.

Page 151. Newspaper headline © 1943, reproduced by permission of the *Des Moines Register.*

Page 158. Cartoon reproduced by permission of the *Buffalo News.*

Page 163. Newspaper headline © 1964, reproduced by permission of the *Des Moines Register.*

Page 164. "Daylight Saving Time" by Louis M. Jones © 1958 (renewed) Unichappell Music Inc. All rights reserved. Used by permission of Warner Bros. Publications U.S. Inc., Miami, Florida 33014.

Page 167. Newspaper headline reproduced by permission of the *St. Paul Pioneer-Press.*

Page 169. Cartoon © 1957, reproduced by permission of the *Pittsburgh Post-Gazette.*

Page 176. Cartoon © 1966, reproduced by permission of the *Des Moines Register.*

Page 194. Cartoon reproduced by permission of the *Blade* (Toledo, OH), 1974.

Page 200. Cartoon reproduced by permission of John L. Hart FLP, and Creators Syndicate, Inc.

Page 215. "Double Daylight Savings Time" written by Bud Davidge © 1987. Song recorded on the album *SIMANI (Bud and Sim) Two for the Show.*

Page 221 Cartoon by Arnie Levin © *The New Yorker* Collection, 1991. All rights reserved.

INDEX